Dendrimers in Medicine and Biotechnology
New Molecular Tools

Dendrimers in Medicine and Biotechnology
New Molecular Tools

U. Boas
Danish Institute of Food and Veterinary Research, Copenhagen, Denmark

J. B. Christensen
Department of General and Organic Chemistry, University of Copenhagen, Copenhagen, Denmark

P. M. H. Heegaard
Danish Institute of Food and Veterinary Research, Copenhagen, Denmark

RSCPublishing

TP 1180
·D45 B63
2006

The front cover image is based on a design from the inside front cover of *Chem Commun.,* 2005, 289–416, from a paper by J. Wu, J. Zhou, F. Qu, P. Bao, Y. Zhang and I. Peng, pg 313.

ISBN-10: 0-85404-852-9
ISBN-13: 978-0-85404-852-6

A catalogue record for this book is available from the British Library

Published by The Royal Society of Chemistry,
Thomas Graham House, Science Park, Milton Road,
Cambridge CB4 0WF, UK

Registered Charity Number 207890

For further information see our web site at www.rsc.org

Typeset by Macmillan India Ltd, Bangalore, India
Printed by Henry Ling Ltd, Dorchester, Dorset, UK

Preface

Dendrimers are a new class of synthetic polymers based on a well-defined cascade motif. These macromolecules may be synthesised to reach the size of nanoobjects having dimensions similar to proteins. Dendrimers allow a highly multivalent presentation of a given molecular motif in a highly defined fashion.

This book concerns the use of this new class of macromolecules in the interdisciplinary field between synthetic organic chemistry, biology, medicine and biotechnology. Dendrimers have initially been applied extensively and studied thoroughly in polymer and material science, but the use of dendrimers as biologically useful compounds, for example in drug delivery or drug construction, is a new research field still in its infancy.

For these new applications of dendrimers, new and important questions need to be answered; how do subtle changes in the molecular motif affect the biological behaviour of a dendrimeric compound comprising multiple copies of that motif? What is the reason some dendrimers are toxic whereas others are not, although their molecular motif is similar?

The ability to construct dendrimers in a highly defined way opens up for the synthetic fine-tuning of molecules fulfilling the desired delicate balance between different biological properties. Furthermore, the ability to construct definite dendrimer architectures opens up for the thorough understanding of the exact nature of interactions taking place between biological entities and a fully synthetic macromolecule.

The present book gives a short and "popular" introduction (Chapter 1) to the dendrimer field, introducing various types of dendrimers, and some general approaches in dendrimer synthesis *e.g.* the divergent and convergent strategy, together with definitions of some common terms (dendrimer generation, shell, *etc.*) applied in this field. Importantly, our definition of dendrimer generations is in accordance with the generation definition initially applied for poly (propylene imine) dendrimers, although the generation numbering of the structurally similar polyamidoamine dendrimers seems to follow another trend. This implies that the generation numbering of the cited literature will follow the definition put out in the book and not in the associated papers. Furthermore, the introducing chapter treats some basic principles in the physicochemical behaviour of some of the common dendrimer classes.

The second chapter covers the interactions between dendrimers and biological systems both *in vitro* and *in vivo*. Important molecular factors affecting the toxicity and biopermeability of these compounds *in vitro* and *in vivo* are described, together with their ability to be transported across membranes (*e.g.* transfection) and tissue barriers (epithelia/endothelia crossing). Biopermeability properties are of crucial importance for the pharmaco-chemical fate of a dendrimer-based drug or drug vehicle.

The subsequent chapters look into the use of dendrimers as drug-delivery devices and drugs. Chapter 3 deals with the development and chemical design of drug transport and delivery vehicles such as host–guest complexes, covalently dendrimer attached drugs (*e.g.* dendrimer prodrugs), self-immolative dendrimer drug systems and targeted drug delivery based on dendrimers.

Chapter 4 concerns the application of dendrimers as drugs and therapeutics in the treatment of antiviral or antibacterial infections as well as for antitumour and anticancer therapy. In addition this chapter goes into the interactions between dendrimers and the immune system, *i.e.* the use of dendrimers as scaffolds in vaccines and/or the use dendrimers as immune-stimulating or immune-suppressing compounds. Just recently dendrimers have been applied in the destabilisation of misfolded prion aggregates responsible for prion-associated diseases *e.g.* Alzheimer's, Diabetes or Mad Cow Disease. These aggregate destabilising dendrimers may constitute an important class of compounds in prion therapeutics and diagnostics.

The final chapter describes the use of dendrimers as mimics for naturally occurring macromolecules or even larger objects, for example, microbial or cellular surfaces, taking advantage of the fact that dendrimers can be synthesised into nanosized structures. As the mode of action of *e.g.* dendrimer antibacterial and antiviral drugs relies on their ability to mimic cellular surfaces of the infected host, there is consequently some "overlap" with Chapter 4 however, in order not to tire the reader, this overlap is kept to a minimum.

Although the main topic of this book concerns dendrimers, the book may apply for polymers having physicochemical properties similar to dendrimers (*e.g.* the larger class of hyperbranched polymers).

Finally, Dr. Ian Law and Kasper Moth-Poulsen are gratefully acknowledged for the proofreading of part of this manuscript.

Good luck!

U. Boas
J.B. Christensen
P.M.H. Heegaard

Table of Contents

**Chapter 1 Dendrimers: Design, Synthesis and Chemical
Properties**   **1**
 1.1 Introduction   1
 1.2 Terms and Nomenclature in Dendrimer Chemistry   3
 1.3 Dendrimer Design   7
 1.4 Dendrimer Synthesis   8
 1.5 Physicochemical Properties of Dendrimers   16
 1.6 Summary   24
 References   24

Chapter 2 Properties of Dendrimers in Biological Systems   **28**
 2.1 Significance of Multivalent Binding in Biological
   Interactions   28
  2.1.1 Dendritic Effect   29
  2.1.2 Carbohydrate Ligands   30
 2.2 Biocompatibility of Dendrimers   32
 2.3 *In Vitro* Cytotoxicity of Dendrimers   32
 2.4 *In Vivo* Cytotoxicity of Dendrimers   43
 2.5 Biopermeability of Dendrimers   46
 2.6 Biodistribution of Dendrimers   55
 2.7 Immunogenicity of Dendrimers   57
 2.8 Summary   57
 References   59

Chapter 3 Dendrimers as Drug Delivery Devices   **62**
 3.1 Introduction   62
 3.2 Dendrimer Hosts   63
  3.2.1 Dendrimer Hosts: Non-Specific
    Interactions with the Dendrimer Core   64
  3.2.2 Dendritic Boxes or Topological
    Trapping of Guests   70

3.2.3   Dendrimer Hosts: Specific Interactions
with the Dendrimer Core   71
3.2.4   Dendrimer Hosts: Non-Polar Interactions
with the Dendrimer Surface Group   72
3.2.5   Dendrimer Hosts: Polar Interactions
with the Dendrimer Surface Groups   73
3.3   Covalently Bound Drug-Dendrimer
Conjugates   76
3.3.1   Self-Immolative Systems   79
3.4   Dendrimers as Gene Transfer Reagents   81
3.5   Summary   85
References   85

Chapter 4   Dendrimer Drugs   90
4.1   Introduction   90
4.2   Antiviral Dendrimers   91
4.3   Antibacterial Dendrimers   97
4.4   Dendrimers in Antitumour Therapy   103
4.5   Dendrimers in Therapy of Other Diseases   106
4.6   Dendrimer-Based Vaccines   107
4.7   Dendrimer Interactions with Proteins.
Solublilisation of Protein Aggregates   119
4.8   Summary   123
References   124

Chapter 5   Dendrimers in Diagnostics   130
5.1   Contrast Agents based on Dendrimers   130
5.1.1   Dendrimer-based Contrast Agents for CT   131
5.1.2   Dendrimer-based Contrast Agents
for MRI   132
5.1.3   Dendrimer-based Contrast Agents
for Scintigraphy   133
5.2   Fluorescence Enhanced by Dendrimers   133
5.3   Dendrimers in Bioassays   135
5.3.1   Dendrimer-Enhanced Signal
Generation   135
5.3.2   Dendrimers for Amplifying the Covalent
Binding Capacity of Solid Phases   139
5.3.3   Microarray Application of Dendrimers   141
5.3.4   Fluorophore-Labelled Dendrimers for
Visualisation Purposes   145

5.3.5 Using Dendrimers to Increase DNA
Extraction Yields   147
5.3.6 Using Dendrimers for direct Detection
of Live Bacteria   148
5.4 Summary   149
References   149

Chapter 6 Dendrimers as Biomimics   **152**
6.1 Introduction   152
6.2 Dendrimers as Protein Mimics   153
6.3 Dendrimers as Artificial Enzymes   161
6.4 Dendrimers as Artificial Antigens, Cell
Surfaces and Antibodies   167
6.5 Summary   170
References   171

Subject Index   **173**

CHAPTER 1

Dendrimers: Design, Synthesis and Chemical Properties

1.1 Introduction

The dendritic structure is a widespread motif in nature often utilised where a particular function needs to be exposed or enhanced. Above ground, trees use dendritic motifs to enhance the exposure of their leaves to the sunlight, which is crucial to maintain life and growth via the photosynthesis. The shade of the tree crown creates a microenvironment maintaining higher humidity and more stable temperatures throughout the day compared to the surroundings. Also beneath ground, the trees have a maximum need to expose a large functional surface when collecting water from the soil. A large dendritic network of roots provides an excellent motif for that purpose (Figure 1.1).

In the "design" of animals and humans, evolution often ends up creating dendritic solutions to enhance particular properties. When breathing air into our lungs the air

Figure 1.1

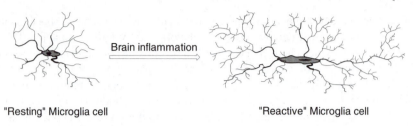

"Resting" Microglia cell   "Reactive" Microglia cell

Figure 1.2 *Activation of a Microglia cell during a pathological state in the brain*

passes through a tremendous dendritic network of bronchioles and alveoli in order to give maximum surface for the transfer of oxygen into the bloodstream. Also the arterial network transporting the oxidised blood to the different organs progress into dendritic patterns, before the blood is transported back to the heart via the venous system.[1] The central nervous system and the brain consist of a large amount of cells growing into dendritic structures in order to gain the largest exchange of material (and information) with the surrounding tissue. Microglia cells serving as multifunctional helper cells in the brain, form dendritic strucures when activated during pathological or degenerative states in the brain (Figure 1.2). Also here the dendritic structure ensures maximum delivery of secreted anti-inflammatory interleukins to the diseased brain tissue.

Another striking example of dendritic structures in nature discovered just recently, is the tremendous number of foot-hairs on the Gecko's feet. These foot-hairs "setae" split up into an impressive dendritic network of tiny foot hairs "spatulae", enabling the Gecko to "stick" to surfaces through dry adhesion without the need of humidity to create surface tension. Examinations of the Gecko's foot-hairs have revealed that the structures of the millions of end foot-hairs are so microscopic that the adhesion between the surface and the gecko foot is thought to be achieved by weak attractive quantum chemical forces from molecules in each foot-hair interacting with molecules of the surface, the so-called Van der Waal forces.[2] By applying a dendritic pattern, the enhancement of a certain function can sometimes greatly exceed the sum of single entities carried on the surface, because of the synergy gained by a dendritic presentation of a function. So nature has, indeed, applied dendritic structures throughout evolution with great success.

In synthetic organic chemistry the creation and design of dendritic compounds is a relatively new field. The first successful attempt to create and design dendritic structures by organic synthesis was carried out by Vögtle and co-workers[3] in 1978. These relatively small molecules were initially named "cascade molecules" and already then Vögtle and co-workers saw the perspectives in using these polymers as, *e.g.* molecular containers for smaller molecules. However, after this first report, several years passed before the field was taken up by Tomalia's group at Dow Chemicals. They had during the years developed a new class of amide containing cascade polymers, which brought these hitherto quite small molecular motifs into well-defined macromolecular dendritic structures. Tomalia and co-workers[4,5] baptised this new class of macromolecules "dendrimers" built up from two Greek words "dendros" meaning "tree" or "branch" and "meros" meaning "part" in Greek. Later refinement and development of synthetic tools enabled the scientists also to synthesise macromolecular structures relying on the original "Vögtle cascade motif".[6,7]

Parallel to polymer chemists taking this new class of compounds into use, dendritic structures also started to emerge in the "biosphere", where J. P. Tam in 1988 developed intriguing dendritic structures based on branched natural amino acid monomers thereby creating macromolecular dendritic peptide structures commonly referred to as "Multiple Antigen Peptide". The Multiple Antigen Peptide is, as we shall see later, a special type of dendrimer.[8]

Dendrimers are also sometimes denoted as "arboroles", "arborescent polymers" or more broadly "hyperbranched polymers", although dendrimers having a well-defined finite molecular structure, should be considered a sub-group of hyperbranched polymers. After the initial reports the papers published on the synthesis, design and uses of dendrimers in chemistry as well as in biological field has had an exponential increase in numbers.[9–14]

1.2 Terms and Nomenclature in Dendrimer Chemistry

Dendrimer chemistry, as other specialised research fields, has its own terms and abbreviations. Furthermore, a more brief structural nomenclature is applied to describe the different chemical events taking place at the dendrimer surface. In the following section a number of terms and abbreviations common in dendrimer chemistry will be explained, and a brief structural nomenclature will be introduced.

Hyperbranched polymers is a term describing a major class of polymers mostly achieved by incoherent polymerisation of AB_n ($n \geq 2$) monomers, often utilising one-pot reactions. Dendrimers having a well-defined finite structure belongs to a special case of hyperbranched polymers (see Figure 1.3). To enhance the availability of dendritic structures, hyperbranched polymers are for some purposes used as dendrimer "mimics", because of their more facile synthesis. However, being polydisperse, these types

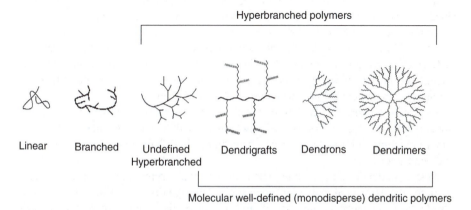

Figure 1.3 *Evolution of polymers towards dendritic structures*

of polymers are not suitable to study chemical phenomena, which generally require a well-defined chemical motif enabling the scientist to analyse the chemical events taking place. The physicochemical properties of the undefined hyperbranched polymers are intermediate between dendrimers and linear polymers.[15]

Dendrigrafts are class of dendritic polymers like dendrimers that can be constructed with a well-defined molecular structure, *i.e.* being monodisperse. However, in contrast to dendrimers, dendrigrafts are centred around a linear polymer chain, to which branches consisting of copolymer chains are attached. These copolymer chains are further modified with other copolymer chains and so on, giving a hyperbranched motif built up by a finite number of combined polymers.[16] Whereas the dendrimer resembles a tree in structure, the core part of a dendrigraft to some extent resembles the structure of a palm-tree.

Dendrons is the term used about a dendritic wedge without a core, the dendrimer can be prepared from assembling two or more dendrons. As we shall see later, dendrons are very useful tools in the synthesis of dendrimers by the segment coupling strategy (convergent synthesis). A class of dendrons, which is commercially available and has been applied with great success in the covalent and non-covalent assembly of dendrimers, are the "Fréchet-type dendrons".[17–19] These are dendritic wedges built up by hyperbranched polybenzylether structure, like the Fréchet-type dendrimers.[17–19] These dendrons have been used in the creation of numerous of dendrimers having different structures and functions.

Generation is common for all dendrimer designs and the hyperbranching when going from the centre of the dendrimer towards the periphery, resulting in homostructural layers between the focal points (branching points). The number of focal points when going from the core towards the dendrimer surface is the generation number (Figure 1.4). That is a dendrimer having five focal points when going from the centre to the periphery is denoted as the 5th generation dendrimer. Here, we abbreviate this term to simply a G5-dendrimer, *e.g.* a 5th generation polypropylene imine and a polyamidoamine dendrimer is abbreviated to a "G5-PPI-" and "G5-PAMAM" dendrimer, respectively. The core part of the dendrimer is sometimes denoted generation "zero", or in the terminology presented here "G0". The core structure thus presents no focal points, as hydrogen substituents are not considered focal points. Thus, in PPI dendrimers, 1,4-diaminobutane represents the G0 core-structure and in PAMAM Starburst dendrimers ammonia represents the G0 core-structure. Intermediates during the dendrimer synthesis are sometimes denoted half-generations, a well-known example is the carboxylic acid-terminated PAMAM dendrimers which, as we shall see later, sometimes have properties preferable to the amino-terminated dendrimers when applied to biological systems.

Shell: The dendrimer shell is the homo-structural spatial segment between the focal points, the "generation space". The "outer shell" is the space between the last outer branching point and the surface. The "inner shells" are generally referred to as the dendrimer interior.

Pincer: In dendrimers, the outer shell consists of a varying number of pincers created by the last focal point before reaching the dendrimer surface. In PPI and PAMAM dendrimers the number of pincers is half the number of surface groups (because in these dendrimers the chain divides into two chains in each focal point).

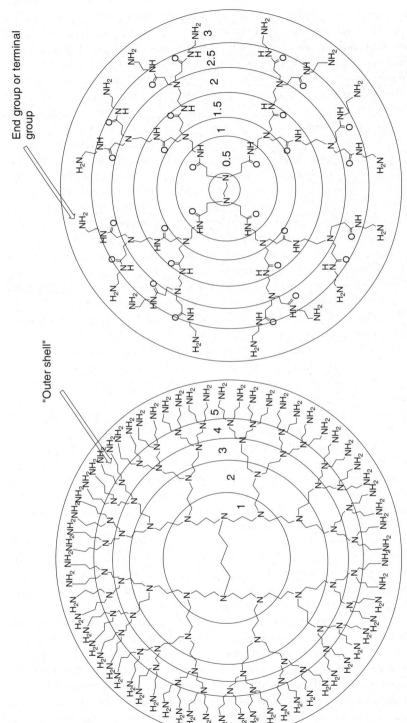

Figure 1.4 *PPI of PAMAM dendrimers with generation of shell depiction*

End-group is also generally referred to as the "terminal group" or the "surface group" of the dendrimer. The word surface group is slightly more inaccurate, in the sense that the dendrimer branches can sometimes fold into the interior of the dendrimer. Dendrimers having amine end-groups are termed "amino-terminated dendrimers".

MAP-dendrimers stand for "Multiple Antigen Peptide", and is a dendron-like molecular construct based upon a polylysine skeleton. Lysine with its alkylamino side-chain serves as a good monomer for the introduction of numerous of branching points. This type of dendrimer was introduced by J. P. Tam in 1988,[8] has predominantly found its use in biological applications, *e.g.* vaccine and diagnostic research. MAP was in its original design a "tree shaped" dendron without a core. However, whole dendrimers have been synthesised based upon this motif either by segmental coupling in solution using dendrons or stepwise by solid-phase synthesis.[20]

PPI-dendrimers stand for "Poly (Propylene Imine)" describing the propyl amine spacer moieties in the oldest known dendrimer type developed initially by Vögtle.[3] These dendrimers are generally poly-alkylamines having primary amines as end-groups, the dendrimer interior consists of numerous of tertiary tris-propylene amines. PPI dendrimers are commercially available up to G5, and has found widespread applications in material science as well as in biology. As an alternative name to PPI, POPAM is sometimes used to describe this class of dendrimers. POPAM stands for POly (Propylene AMine) which closely resembles the PPI abbreviation. In addition, these dendrimers are also sometimes denoted "DAB-dendrimers" where DAB refers to the core structure which is usually based on DiAminoButane.

PEI-dendrimers is a less common sub-class of PPI dendrimers based on Poly (Ethylene Imine) dendritic branches. The core structure in these dendrimers are diamino ethane or diamino propane.

PAMAM-dendrimers stand for PolyAMido-AMine, and refers to one of the original dendrimer types built up by polyamide branches with tertiary amines as focal points. After the initial report by Tomalia and co-workers[4,5] in the mid-1980s PAMAM dendrimers have, as the PPI dendrimers, found wide use in science. PAMAM dendrimers are commercially available, usually as methanol solutions. The PAMAM dendrimers can be obtained having terminal or surface amino groups (full generations) or carboxylic acid groups (half-generations). PAMAM dendrimers are commercially available up to generation 10.[17]

Starburst dendrimers is applied as a trademark name for a sub-class of PAMAM dendrimers based on a tris-aminoethylene-imine core. The name refers to the star-like pattern observed when looking at the structure of the high-generation dendrimers of this type in two-dimensions. These dendrimers are usually known under the abbreviation PAMAM (Starburst) or just Starburst.

Fréchet-type dendrimers is a more recent type of dendrimer developed by Hawker and Fréchet[17–19] based on a poly-benzylether hyperbranched skeleton. This type of dendrimer can be symmetric or built up asymmetrically consisting of 2 or 3 parts of segmental elements (dendrons) with, *e.g.* different generation or surface motif. These dendrimers usually have carboxylic acid groups as surface groups, serving as a good anchoring point for further surface functionalisation, and as polar surface groups to increase the solubility of this hydrophobic dendrimer type in polar solvents or aqueous media.

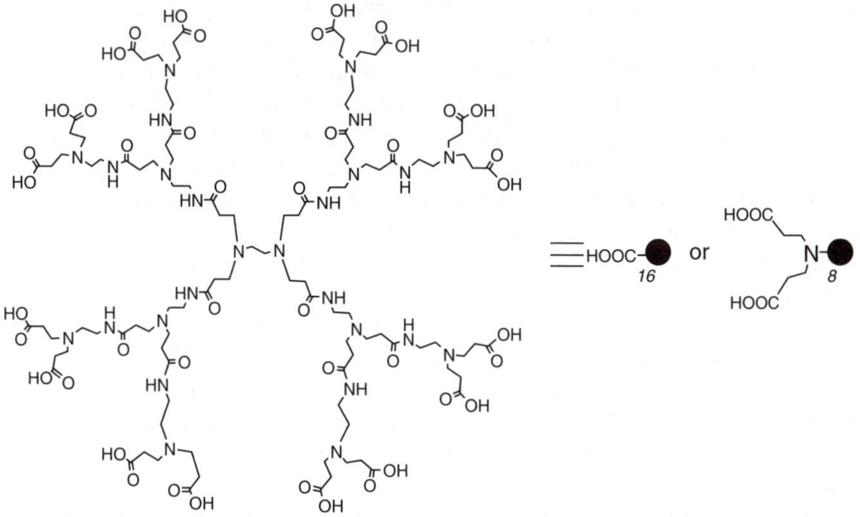

Figure 1.5 *"Black ball" symbol for a 2.5 G-PAMAM dendrimer*

"Black ball" nomenclature: Because of the large molecular structure of a dendrimer, the full picture of, *e.g.* reactions taking place on the dendrimer surface or in the outer shell can be difficult to depict. A way to facilitate the depiction of these macromolecules is by showing the inner (and unmodified) part of the dendrimer as a "black ball". Depending on whether the reaction takes place at the surface groups or in the outer shell, the appropriate part of the molecular motif, *e.g.* the outer pincers, may be fully drawn out to give a concise picture of a reaction involving the outer shell (see Figure 1.5). In this way the picture of reactions taking place at the dendrimer surface or in the outer shell is greatly simplified.

1.3 Dendrimer Design

After the initial reports and development of these unique well-defined structures, chemists have begun to develop an excessive number of different designs of dendrimers for a wide variety of applications. Newkome and co-workers[22] developed the unimolecular micelle consisting of an almost pure hydrocarbon scaffold, Majoral and Caminade introduced the multivalent phosphorus to create intriguing new dendrimeric designs and dendrimers having new properties. Other third period elements like silicon and sulfur have been implemented in the dendritic structures resulting in dendrimers having properties quite different from the classical PAMAM and PPI designs.[23] The monomers applied in the build-up of a dendrimer are generally based on pure synthetic monomers having alkyl or aromatic moieties, but biological relevant molecules like carbohydrates,[24] amino acids[20] and nucleotides[25-27] have been applied as monomers as well (Figure 1.6).

Using biological relevant monomers as building blocks presents an intriguing opportunity to incorporate biological recognition properties into the dendrimer.[20,24]

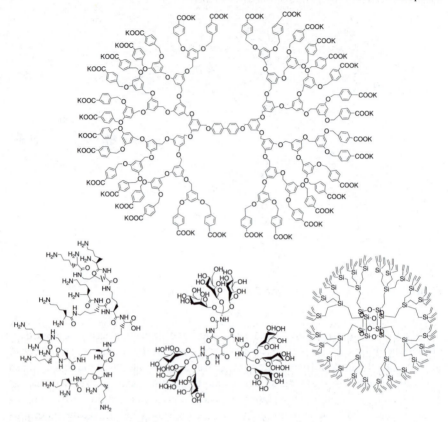

Figure 1.6 *Different dendrimer designs. Top: G3-Fréchet-type dendrimer. Bottom from the right: MAP dendron, glycodendrimer and a silicon-based dendrimer*

As we shall see, also metal ions serve as good focal points and have found extensive use in various functional dendrimer designs as well as in the synthesis of dendrimers by self-assembly.[28]

1.4 Dendrimer Synthesis

Divergent dendrimer synthesis: In the early years of dendrimers, the synthetic approach to synthesise the two major dendrimer designs, the PPI and PAMAM, relied on a stepwise "divergent" strategy. In the divergent approach, the construction of the dendrimer takes place in a stepwise manner starting from the core and building up the molecule towards the periphery using two basic operations (1) coupling of the monomer and (2) deprotection or transformation of the monomer end-group to create a new reactive surface functionality and then coupling of a new monomer *etc.*, in a manner, somewhat similar to that known from solid-phase synthesis of peptides or oligonucleotides.

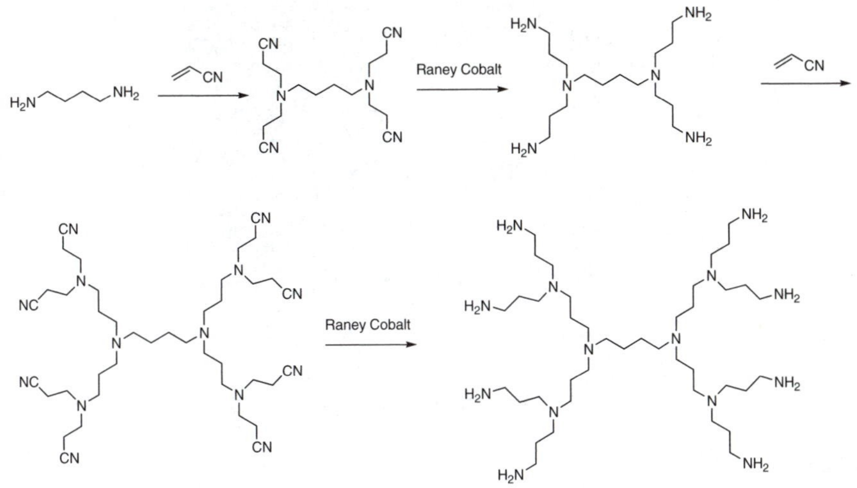

Figure 1.7 *Poly(propylene imine) dendrimer synthesis by divergent strategy*

For the poly (propyleneimine) dendrimers, which are based on a skeleton of poly alkylamines, where each nitrogen atom serves as a branching point, the synthetic basic operations consist of repeated double alkylation of the amines with acrylonitrile by "Michael addition" results in a branched alkyl chain structure. Subsequent reduction yields a new set of primary amines, which may then be double alkylated to provide further branching *etc.* (Figure 1.7)[7]

PAMAM dendrimers being based on a dendritic mixed structure of tertiary alkylamines as branching points and secondary amides as chain extension points was synthesised by Michael alkylation of the amine with acrylic acid methyl ester to yield a tertiary amine as the branching point followed by aminolysis of the resulting methyl ester by ethylene diamine.

The divergent synthesis was initially applied extensively in the synthesis of PPI and PAMAM dendrimers, but has also found wide use in the synthesis of dendrimers having other structural designs, *e.g.* dendrimers containing third period heteroatoms such as silicium and phosphorous.[23] Divergent synthesis of dendrimers consisting of nucleotide building blocks has been reported by Hudson and co-workers.[25] The divergent stepwise approach in the synthesis of nucleotide dendrimers and dendrons is interesting from a biochemical perspective as it may mimic the synthesis of naturally occuring *lariat* and *forked introns* in microbiology.[25]

To discriminate between the divergent build-up of a linear molecule, *e.g.* a peptide/protein in a stepwise manner, and the proliferating build-up of a dendrimer also by a divergent methodology, Tomalia and co-workers have applied the term "Amplified Geneologically Directed Synthesis" or A-GDS to describe divergent dendritic synthesis, as an opposite to a "Linear Geneologically Directed Synthesis" (L-GDS) performed in, *e.g.* Merrifield solid-phase peptide synthesis (Figure 1.8).[29]

There are two major problems when dealing with divergent synthesis of dendrimers, (1) the number of reaction points increases rapidly throughout the synthesis

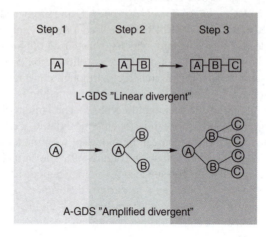

Figure 1.8

of the dendrimer, starting with 2 points in a G0-PPI dendrimer and ending up with 64 for G5-PPI dendrimer. This rapid increase in number of end-groups to be functionalised, combined with the following rapid increase in molecular weight resulting in slower reaction kinetics, makes the synthesis of the dendritic network to create higher generation dendrimers increasingly difficult, even when using high yielding reactions. Therefore the divergent approach may lead to increasing deletions throughout the growth of the dendrimer, resulting in numerous of defects in the higher generation dendrimer product. The synthesis of PPI dendrimers has to some extent been hampered by the creation of defects throughout the synthesis of higher generation dendrimers where it has been shown that the content of molecular perfect G5-PPI dendrimer in the product is only approximately 30%.[10] In the case of PPI dendrimers, the divergent approach is applied with most success in the synthesis of lower generation dendrimers (that is dendrimers upto G3). In case of the PPI dendrimers defects may also emerge in the final high generation product after synthesis, as a result of the PPI-structure being based on short spacer monomers. This creates an increasing molecularly crowded structure throughout the generations, leading to the loss of dendrimer branches and wedges because of increased susceptibility to, *e.g.* β-elimination reactions. Secondly, when performing divergent synthesis it is hard to separate the desired product from reactants or "deletion products", because of the great molecular similarity between these by-products and the desired product. Despite these drawbacks being observed predominantly in the synthesis of high generation PPI dendrimers, the divergent approach has been applied in the synthesis a large variety of different dendrimer designs with great success.

Generally the divergent approach leads to the synthesis of highly symmetric dendrimer molecules, however, recently scientists have taken up the possibility to create heterogeneously functionalised dendrimers by the divergent approach, leading to dendrimers having several types of functionalities bound to the surface.[30,31] This field is an exiting opportunity to use conventional dendrimers as scaffolds for different molecular functions.

Divergent dendrimer synthesis by self-assembly: Using biological building blocks gives a high degree of recognition, which can be used for highly specific self-assembly of the building blocks. Nilsen and co-workers[27] have used oligonucleotide building blocks in divergent self-assembly of dendrimers, followed by cross-binding to stabilise the self-assembled dendrimer construct. Although constituting an interesting example of a divergent segment-based synthesis, this method is quite complex due to the complex structure of the each building block. The building blocks consist of two annealed oligonucleotides annealing together at the mid-section, thus dividing out into four arms, which then each can be modified with monomers having complementary motifs on their arms and so on (Figure 1.9). The surface monomers may be modified with, *e.g.* a labelling group also by oligonucleotide annealing. Although this synthetic method does not result in perfect dendrimers, it still provides an intriguing alternative to divergent dendrimer synthesis relying on more traditional low-molecular monomers. This dendrimer design has been applied as scaffolds for biomolecules in diagnostics (see Chapter 5).

Convergent dendrimer synthesis: Segment coupling strategies began to be applied in peptide synthesis to circumvent the increasingly low reactivity experienced in stepwise divergent synthesis of large oligopeptides on solid-phase. With this new approach, peptide synthesis was taken a step further towards pure chemical synthesis of high molecular weight polypeptides and proteins (for a survey on convergent peptide/protein systhesis, see Ref. 32). This segmental coupling or convergent strategy also found its way into the creation of dendritic macromolecules, first implemented by Hawker and Fréchet[17–19] in their synthesis of poly-benzylether containing dendrimers which gave highly monodisperse dendrimer structures. A powerful alternative to the divergent

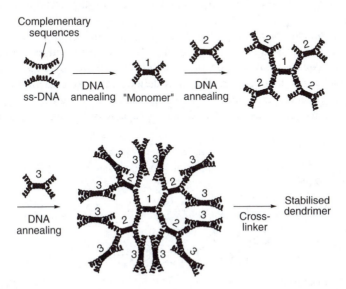

Figure 1.9 *Schematic depiction of divergent dendrimer synthesis based on oligonucleotode monomers using a self-assembly by oligonucleotide annealing, and after the final product is assembled, consolidation of the self-assembly by crosslinking*

approach had been introduced and this new tool was promptly taken up among other synthetic chemists working in the dendrimer field.

In contrast to the divergent method, the convergent method construct a dendrimer so to speak from the surface and inwards towards the core, by mostly "one to one" coupling of monomers thereby creating dendritic segments, dendrons, of increasing size as the synthesis progress. In this way the number of reactive sites during the proliferation process remains minimal leading to faster reaction rates and yields. Another advantage of this methodology is the large "molecular difference" between the reactant molecule and the product, facilitating the separation of the reactants from the product during the purification process. The final part of the convergent synthesis ends up at the core, where two or more dendritic segments (dendrons) are joined together, creating the dendrimer, the convergent strategy thus generally has an inverse propagation compared to the divergent strategy (Figure 1.10).

In addition, the convergent strategy is an obvious tool in the synthesis of asymmetric dendrimers, or dendrimers having mixed structural elements, where instead of coupling two equal segments in the final segment coupling reaction(s), different segments are coupled together to create dendrimers with heterogeneous morphologies.[33] This relatively easy approach to create heterogeneous dendrimers opens up to intriguing fields of incorporating several "active sites" in one dendrimer to create multifunctional macromolecular structures.

After its advent, the convergent strategy has also been used for the synthesis of a great variety of dendrimers having different core functionalities, where the core is introduced in the final step and modified with dendrons to create the complete dendrimer. This methodology facilitates the synthesis of dendrimers with different core functions, *e.g.* for fluorescence labelling or for the creation of artificial enzymes.

Convergent dendrimer synthesis by self-assembly: Much effort has been given to build up dendrimers in a non-covalent manner by convergent self-assembly of dendrons. The dendrons may contain functionalities capable of hydrogen bonding or metal complex bonding *etc.*, creating well-defined complexes having dendrimeric structures. The area was initially explored by Zimmerman's group[34] who built up dendrimers through self-assembly of dendrons capable of hydrogen bonding.

Divergent   Convergent

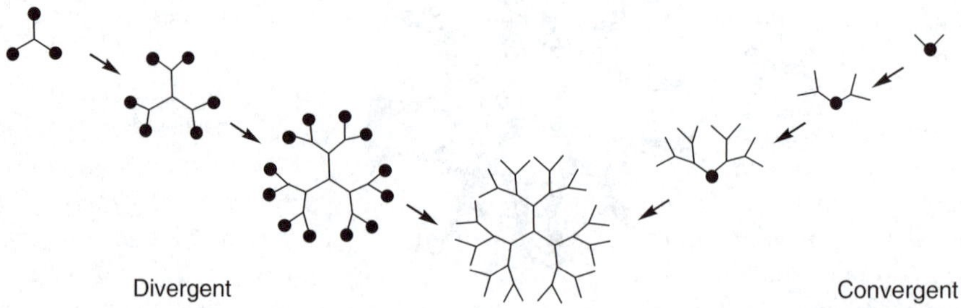

Figure 1.10 *Divergent versus convergent strategy, The black dots mark the "functionalising sites"*

When utilising hydrogen bonding as the "glue" to bind the dendrimer together, the general requirement is that the hydrogen bonding units chosen form complexes which are stable enough to be isolated. When designing these self-assembled products for biological applications as, *e.g.* drug delivery, or self-assembled drugs the hydrogen bonding keeping the segments together should be stable under highly polar physiological conditions, *i.e.* buffered aqueous media containing a high concentration of ions.

In the earliest attempts of non-covalent synthesis of dendrimers, Zimmerman's group[34] applied Fréchet dendrons containing bis-isophtalic acid, which in chloroform spontaneously formed hexameric aggregates through carboxylic acid–carboxylic acid hydrogen bonding. These hexameric aggregates were stable in apolar solvent like chloroform, but dissociated in more polar solvents like tetrahydrofuran and dimethyl sulfoxid, in which NMR only showed the existence of the corresponding monomers.

Another early report on the synthesis of dendrimers utilising self-assembly of hydrogen bonding dendrons was launched by Fréchet's group[35] who applied dendrons with complementary melamine and cyanuric acid functionalities for hydrogen bonding. These dendrons formed hexameric aggregates in apolar solvents, but as the "Zimmerman dendrons" these assemblies dissociated upon exposure to polar solvents. In order to increase the stability of self-assembled dendrimers in polar solvents, Zimmerman and his group[36] utilised the ureidodeazapterin moieties capable of forming exceptional strong hydrogen bonds, Frechet-type dendrons bound to this group via a spacer hydrogen bonded together and gave dimeric up to hexameric aggregates which had high stability both in apolar solvents like chloroform and in water.

An alternative to the development and design of synthetic molecular motifs capable of molecular recognition is the use of nature's own molecular motifs for highly specific molecular recognition. Single strand DNA (ss-DNA) forming stable complexes upon annealing with a complementary single DNA strands to form a DNA duplex has been applied as recognition motifs and bound to dendritic wedges. In this way two "complementary" dendrons each carrying one DNA-strand could be coupled together with high specificity forming a bi-dendronic dendrimer.[37]

A similar idea has been applied in the synthesis of supramolecular drugs for tumour targeting based on a "bi-dendrimer" by duplex formation of two differently functionalised dendrimers each containing a complementary oligonucleotide sequence (Figure 1.11)[38]

Metal ions with their Lewis acid properties may serve as good acceptors for appropriate electron pair donors attached to the dendrons, thus using the metal ion as the core for assembly of dendronic ligands. Dendrimers assembled around a lanthanide metal (*e.g.* Europium) have been created by Kawa and Fréchet.[39] The metal being in the core of the dendrimer experiences a microenvironment kept away from interacting with the surroundings, resulting in enhanced photoluminiscence. The site isolation retards energy transfer processes with the surroundings as well as the formation of metal clusters, which leads to the quenching observed in small ligand complexes (*e.g.* triacetates) of these elements.[39] Narayanan and

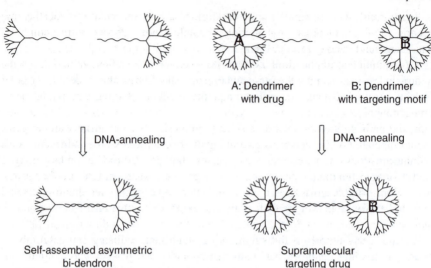

A: Dendrimer   B: Dendrimer
with drug   with targeting motif

DNA-annealing   DNA-annealing

Self-assembled asymmetric   Supramolecular
bi-dendron   targeting drug

Figure 1.11 *Left: Schematic depiction of convergent dendron self-assembly creating asymmetric dendrimer constructs. Right: The use of ss-DNA in highly specific construction of multifunctional dendritic drugs*

Wiener[40] assembled dendrons around a Co^{3+} ion by formation of an octahedral complex where the metal was surrounded by three bidentate dendrons spreading out into six dendrimer branches. Cobalt is an extremely interesting element because of the large difference in properties when going from Co^{2+} generally forming quite unstable complexes with mostly tetrahedral symmetry compared to Co^{3+} which forms stable octahedral complexes. The use of transition metals as templates for dendrimer assembly presents the possibility to oxidise or reduce the metal centre, which may result in a new conformation or altered stability of the assembly, thereby creating a material responsive to oxidation or reduction from the surroundings.

An exciting aspect when applying weak binding forces compared to traditional covalent assembly, is the observation that even small molecular changes (or defects) in the respective monomers may have a strong effect on the ability for the final non-covalent dendrimer product to form.[41] In that sense this methodology closely resembles "natures way" of building up macromolecular structures, where even small "mutations" in, *e.g.* the amino acid side-chain motifs may lead to catastrophic consequences on the three-dimensional shape of the final protein and disable a particular biological function of that protein. The field of creating macroscopic dendrimeric nano-objects by self-assembly is a very important research area in order to get a closer understanding of the factors governing self-assembly processes, *e.g.* the molecular information concerning the shape of the final supramolecular product carried by the respective monomers. Furthermore, a deeper knowledge opens up to create molecular structures, which can change morphology and function upon different stimuli. The non-covalent methodology is a very important approach for the creation of, *e.g.* functional biomaterials, capable of responding to the complex processes found in biological systems.

Self-assembly has been combined with conventional covalent synthesis of dendrimers by Shinkai and co-workers,[42] who used self-assembly of the dendrimers as templates for subsequent consolidation of the product by cross-linking the dendrimer together. In organic chemistry, the self-assembly processes leading to the right supramolecular product is at present time a relatively straightforward process due to the relatively simple supramolecular patterns and highly ordered structures of the building blocks. In nature, however, the self-assembly/consolidation process is a highly complicated matter, *e.g.* in the refolding of proteins from a denatured state. Going from a highly disordered denatured state to a highly ordered native state is a highly unfavourable process with respect to entropy (Figure 1.12).

In order to fold or refold the protein sequences into three-dimensional protein structures the unfolded or partially unfolded protein is taken up by a class of proteins called "Chaperones". Chaperones are cytoplasmic proteins that serve as templates in the folding process to give the final, and biological functional protein, and in preventing aggregate formation due to intermolecular hydrophobic interactions. The chaperones are also denoted "Heat Shock Proteins" because of their ability to prevent denaturation of proteins, which otherwise would be lethal, when our organisms are subjected to fever during illness.[43]

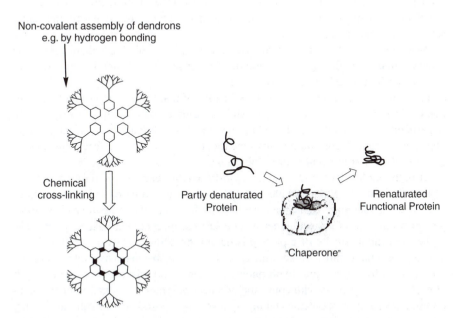

Figure 1.12 *Schematic depiction of, left: chemical assembly of a dendrimer via a self-assembly/in situ cross-binding strategy in comparison to the complex folding process of proteins in nature being mediated by Chaperones (right)*

1.5   Physicochemical Properties of Dendrimers

As the dendrimer grows, the different compartments of the dendritic structure begin to show distinct features which are amplified with increasing generation. The dendrimer structure may be divided into three parts:

- A multivalent surface, with a high number of functionalities. Dependent on the dendrimer generation, the surface may act as a borderline shielding off the dendrimer interior from the surroundings. This increasingly "closed" surface structure may result in reduced diffusion of solvent molecules into the dendrimer interior.
- The outer shell, which have a well-defined microenvironment, to some extent shielded from the surroundings by the dendrimer surface. The very high number of functionalities located on the surface and the outer shell are well-suited for host–guest interactions and catalysis where the close proximity of the functional motifs is important.
- The core, which as the dendrimer generation increases, gets increasingly shielded off from the surroundings by the dendritic wedges. The interior of the dendrimer creates a microenvironment which may have very different properties compared to the surroundings. For example as decribed elsewhere, water-soluble dendrimers with an apolar interior have been constructed to carry hydrophobic drugs in the bloodstream.[44]

The three parts of the dendrimer can specifically be tailored towards a desired molecular property or function of the dendrimer such as drug delivery, molecular sensors, enzyme mimics, *etc.*

When looking at the molecular size and properties of dendrimers, one soon observes that the molecular dimension of a higher generations dendrimer is comparable to medium-sized proteins (Table 1.1).[14]

Therefore, it was already early in the history of dendrimers suggested that these nanoscale polymers would serve as synthetic mimics of proteins.[45] However, the hyperbranched structure of the dendrimer creates a highly multivalent surface, exposing a much higher number of functional groups on the surface compared to proteins of similar molecular size (Table 1.1).

Also, the molecular weight of, *e.g.* a G6-PAMAM dendrimer is only around half of that of a protein of comparable molecular size (*e.g.* ovalbumin). This is a consequence of the fact that a dendrimer, because of the molecular structure (tree shaped) generally has a lower molecular density, *i.e.* less compact compared to a protein. The higher molecular density of a protein is due to the ability to tightly fold the linear polypeptide chain into a three-dimensional structure by extensive intramolecular ion-pairing, hydrogen and hydrophobic bonding and disulfide cross-binding.[46] However, in comparison with conventional linear polymers, the dendrimers are generally more compact molecules taking up a smaller hydrodynamic volume.[47] X-ray analysis on supramolecular dendrimer aggregates has revealed that the molecular shape of the dendrimer upon increasing generation becomes increasingly globular (*i.e.* more spherical in contrast to linear shaped), in order to spread out the larger molecular structure with a minimal repulsion between the segments.[48]

Table 1.1 *Physicochemical properties of dendrimers in comparison to various biological entities*

Type of molecule	Molecular weight	pI/surface charge	Diameter	Number and type of surface functional groups*
G3-PAMAM (Starburst†)	2411	/+	2.2 nm	12 primary amines
G6-PAMAM‡	28.788	11/+	6.5 nm	128 primary amines
G6-PAMAM-OH	28.913	9/0	—	128 hydroxyls
Medium sized protein (ovalbumin)	43.000	5/+ and –	5 nm	20 primary amines 10 phenol groups 4 thiols, 7 imidazoles
Large protein (Keyhole Limpet Hemocyanin)	~5.000.000	/+ and –	—	Approximately 2000 primary amines, 700 thiols, 1900 phenols
Virus	~40.000.000	—	50–200 nm	—
Prokaryotic bacteria	—	Mainly negative	1–2 μm (30 nm cell membrane and cell wall)	—
Eukaryotic cell	—	Mainly negative	20 μm (9 nm cell membrane)	—

*Protein functional groups not necessarily surface localised, †core group is trifunctional, branches are made up of ammonia and ethylenediamine building blocks; Starburst is a Trademark of Dendritech Inc., Midland, MI, US, ‡core group is tetrafunctionalised, branches are made up of methyl acrylate and ethylenediamine building blocks.

The use of dendrimers as protein mimics has encouraged scientist to carry out studies to investigate the physicochemical properties of dendrimers in comparison to proteins. Being nano sized structures, dendrimers may respond to stimuli from the surroundings and can, like proteins, adapt a tight-packed conformation ("native") or an extended ("denaturated") conformation, depending on solvent, pH, ionic strength and temperature. However, there are some major differences in the molecular structures of dendrimers in comparison to proteins, resulting in a different physicochemical response of a dendrimer compared to a protein. The dendrimer architecture incorporates a high degree of conjunction consisting of a network of covalent bonds, which results in a somewhat less flexible structure than found in proteins.

Numerous of studies have been carried out to investigate the physicochemical properties of dendrimers applying computer simulations and chemical analytical techniques. And in order to optimise the computer models to give a realistic picture, a large amount of comparative studies have been carried out between predictions-based theoretical calculations and experimental results by chemical analysis.[49,50]

Dendrimers and the effect of molecular growth: The conformational behaviour of a dendrimer upon growing to higher generations are determined by (1) the molecular dimensions of the monomers–short monomers induce rapid proliferation of chains within a small space (2) the flexibility of the dendrons and (3) the ability of the end-groups to interact with each other, *e.g.* by hydrogen bonding creating a dense outer shell.

An initial attempt to predict the intramolecular behaviour of a dendrimer upon increasing the generation number using molecular simulations was reported by the French scientists De Gennes and Hervet,[51] who already in 1983 presented a modification of the "Edwards self-consistent field" theory to describe the conformational characteristics upon growth of a PAMAM (Starburst) dendrimer. Their analyses concluded that upon growth, the periphery (outer shell) of the dendrimer becomes increasingly crowded whereas the molecular density of the core region remains low throughout the molecular growth. As no back-folding (dendrons folding into the interior of the dendrimer) is taken into account, the increasing molecular crowding in the outer shell will give a limitation on the generation number that a starburst dendrimer can grow to.

One major problem in applying this model for dendrimers having, *e.g.* amine surface groups is that it does not take into account that the dendrons in these compounds have a relatively high mobility because of the lack of binding interactions between both the dendrimer arms and the functionalities at the surface. This larger mobility enables the dendrons to fold inwards towards the dendrimer interior as a consequence of entropy, disfavouring the more ordered De Gennes dense shell packing conformation.[49] Thus, the structural behaviour of the dendrimer upon growing to higher generations is determined by the ability of the surface functionalities to form a network with each other via, *e.g.* hydrogen bonding or ion pairing thereby consolidating a dense outer shell. For this reason, the "De Gennes model" has generally been opposed as a suitable model to describe unmodified flexible dendrimers as, *e.g.* amino-terminated PPI and PAMAM dendrimers.[50] However, in cases where the dendrimer contain surface groups capable of hydrogen bonding a dendrimeric motif with a very dense periphery (outer shell) and a hollow core may be obtained. An example of "dense-shell behaviour" has been investigated by Meijers group[52] who modified the surface amino groups of high-generation PPI dendrimers with Boc-phenyl alanine. Boc-phenyl alanine formed numerous of hydrogen bonds between the outer shell amides achieved by the amidation of the dendrimer. In case of the G5-PPI dendrimer, an outer shell was obtained with such a high molecular density that small molecules, *e.g.* Rose Bengal and *para*-nitrobenzoic acid could be entrapped inside the dendrimer without leakage to the surrounding solvent. This dense shell dendrimer was named "the dendritic box", and was besides being seminal in understanding fundamental structural chemistry of dendrimers, the first experimental report pointing towards using dendrimers as molecular containers, for *e.g.* drug delivery (see Chapter 3). Also, later studies of PPI dendrimers modified with amino acids capable of forming hydrogen bonds did show a good correlation with De Gennes "dense shell packing model" when increasing the generation number for these systems.[53] In this and similar cases the dense shell model of De Gennes and Hervet is followed, because the hydrogen bonding between the end-groups disfavour back-folding, which would otherwise lead to a higher molecular density in the interior of the dendrimer (Figure 1.13).

In order to give a more realistic picture on the molecular density in dendrimers having a more flexible structure Lescanet and Muthukumar[54] used "kinetic growth" simulations to predict the molecular conformation of the Starburst molecules. Using this approach they found that extensive back-folding may be found at the late stages

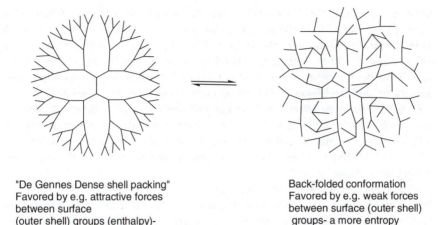

"De Gennes Dense shell packing"
Favored by e.g. attractive forces
between surface
(outer shell) groups (enthalpy)-
highly ordered state

Back-folded conformation
Favored by e.g. weak forces
between surface (outer shell)
groups- a more entropy
driven, less ordered state

Figure 1.13 *Schematic depiction of the consequence of back-folding resulting in an increased molecular density in the interior of a dendrimer*

of dendritic growth. These predictions were confirmed by experimental observations performed on unmodified Starburst PAMAM amino-terminated dendrimers (G0–G7) using ^{2}H and ^{13}C NMR. By using the NMR correlation- and spin-lattice relaxation times, the mobility of the dendritic segments (dendrons) upon increasing generation could be measured. The carbon NMR experiments revealed no supression of mobility of the dendritic chain ends (termini), thus a low mobility of the chain ends is a condition for dense packing of functional groups on the dendrimer surface (*e.g.* De Gennes). In addition, increased average correlation times (τ) for the interior segments, indicated an increasing molecular density in the interior as a result of back-folding.[55] ^{2}H-NMR relaxation experiments, to study chain mobility, indicated a less restricted (faster) segmental motion of the chain ends (opposing the model of De Gennes) in comparison to the chains of the interior of the dendrimer.[56] These findings were in accordance with the molecular simulations reported by Lescanet and Muthukumar, approving this model to describe these types of dendrimers. Also, calculations based on molecular dynamics indicate that flexible dendrimers of all generations exhibit a dense core region and a less dense plateau region close to the periphery of the molecule, *i.e.* low generation dendrimers have conformations with low degree of back-folding ("density overlap") compared to higher generations. Upon reaching higher generations, the amount of back-folding increases upto the G8 dendrimers, where the molecular density is nearly uniform over the entire dendrimer.[57]

Comparative studies have been carried out to determine the shape and evaluate the change in steric interactions in amino-terminated PAMAM dendrimers compared to carbosilane dendrimers upon increasing generation. The steric repulsion is determined by the "scaled steric energy parameter". Carbosilane dendrimers are more spherical in shape compared to PAMAM with the smaller generation dendrimers being less spherical than the higher generation dendrimers. As carbosilane dendrimers are more spherical, the higher generation dendrimers are capable of having an increased number of terminal groups on the molecular surface without increase of molecular density in the outer shell region. This may be due to silicon, being a third period element with a more flexible bond geometry. For PAMAM dendrimers, the steric repulsion becomes almost constant with G>4, whereas for carbosilane dendrimers the steric repulsion decreases upon increasing generation number.[58]

Dendrimers and the effect of pH: Amino-terminated PPI and PAMAM dendrimers have basic surface groups as well as a basic interior. For these types of dendrimers with interiors containing tertiary amines, the low pH region generally leads to extended conformations due to electrostatic repulsion between the positively charged ammonium groups.

Applying molecular dynamics to predict the structural behaviour of PAMAM dendrimers as a function of pH show that the dendrimer has an extended conformation, based on a highly ordered structure at low pH (pH≤ 4). At this pH, the interior is getting increasingly "hollow" as the generation number increases as a result of repulsion between the positively charged amines both at the dendrimer surface and the tertiary amines in the interior.

At neutral pH, back-folding occurs which may be a consequence of hydrogen bonding between the uncharged tertiary amines in the interior and the positively charged surface amines. At higher pH (pH≥ 10) the dendrimer contract as the charge of the molecule becomes neutral, aquiring a more spherical (globular) structure based on a loose compact network, where the repulsive forces between the dendrimer arms and between the surface groups reaches a minimum.[59] At this pH, the conformation has a higher degree of back-folding as a consequence of the weak "inter-dendron" repulsive forces (Figure 1.14).

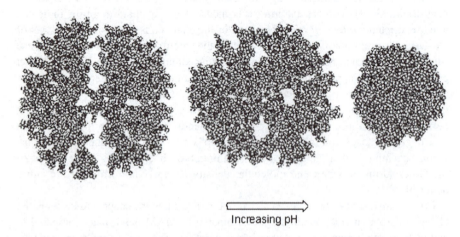

Increasing pH

Figure 1.14 *Three-dimensional structure of a G6-PAMAM dendrimer, under different pH. Calculations is based on molecular dynamics*
(Reprinted from *Macromolecules*, 2002, **35**, 4510, with permission. ©2002, American Chemical Society)

Calculations as well as experimental data generally conclude that dendrimers (G5–G7) are conformationally more affected by change in pH and ionic strength in comparison to higher generation dendrimers (*e.g.* G8). The reason for this may be found in the somewhat more restricted motion of the outer shell chain segments in the higher generation dendrimers, leading to a more globular-shaped molecule despite different conditions in the surroundings.[60] As a curiosum, recent investigations show that amino-terminated PAMAM and PPI dendrimers in addition to their pH dependent conformational changes also fluoresce at low pH.[61]

When looking at the pH-dependent conformational changes of PPI dendrimers having acidic (carboxylic acid) end-groups, the picture is somewhat different compared to what is observed for their amino-terminated counterparts (Figure 1.15). Small angle neutron scattering (SANS) and NMR measurements of self-diffusion coefficients at different pH values show that at pH 2 the dendrimer core has the most extended conformation due to the electrostatic repulsion between the positively charged protonated tertiary amines, leading to a large radius of the core, whereas the dendrimer reaches its minimum radius at pH 6, where the amount of positively charged amines equals the amount of negatively charged carboxylic groups (isoelectric point) resulting in a "dense core" conformation more subjective to back-folding. Thus, at pH 6 some degree of back-folding occurs as a result of attractive Coulomb interactions between the negatively charged surface carboxy-groups and the positively charged tertiary amines in the inner shells of the dendrimer.[62] This shows that back-folding is not only a result of weak forces leading to a uniform molecular density of the dendrimer (entropy), but may also be mediated by attractive forces (enthalpy) between inner parts of the dendrons and surface groups. In the carboxy-PPI dendrimers a back-folded conformation minimise the repulsion between the negatively charged surface groups and between the positively charged inner shell amines leading to a lower repulsive energy of the system. At pH 11 the electrostatic repulsion between the negative charged forces the surface groups apart to give a more extended conformation with a highly expanded surface area (Figure 1.15).

Dendrimers and the effect of solvent: The ability of the solvent to solvate the dendrimer structure is a very important parameter when investigating the conformational state of a dendrimer. Molecular dynamics has been applied to study the variation of

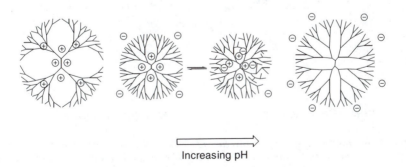

Increasing pH

Figure 1.15 *Two-dimensional depiction of conformational changes upon different pH of a carboxy-terminated PPI-dendrimer*

dendrimer conformation as a function of dendrimer generation in different solvents.[57] Dendrimers of all generations generally all experience a larger extend of back-folding with decreasing solvent quality, *i.e.* decreasing solvation. However, being more flexible, the low generation dendrimers show the highest tendency towards back-folding as a result of poor solvation compared to the higher generation dendrimers.

NMR studies performed on PPI dendrimers conclude that an apolar solvent like benzene, poorly solvates the dendrons favouring intramolecular interactions between the dendrimer segments and back-folding. However, a weakly acidic solvent like chloroform can act as a hydrogen donor for the interior amines in a basic dendrimer like PPI, leading to an extended conformation of the dendrimer because of extensive hydrogen bonding between the solvent and the dendrimer amines.[63] Both experimental as well as theoretical studies on amino-terminated PPI and PAMAM dendrimers (polar dendrimers) show the tendency that apolar aprotic ("poor") solvents induce higher molecular densities in the core region as a result of back-folding, whereas polar ("good") solvents solvate the dendrimer arms and induce a higher molecular density on the dendrimer surface.

Interestingly, dendrimers having polar surface groups to some extent resemble proteins in their conformational behaviour when subjecting these structures to more apolar conditions, in the sense that back-folding of the polar surface groups may expose the more hydrophobic dendrimer parts to the surroundings leading to a decreased surface polarity of the back-folded dendrimer. A similar behaviour has been observed in the adsorption of proteins onto hydrophobic surfaces, giving a highly denatured (unfolded) state of the protein exposing its interior hydrophobic regions to interact with the surface (Figure 1.16).[64]

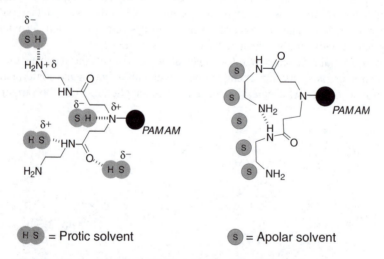

Figure 1.16 *Proposed scheme for solvation of a dendrimer under different solvent conditions. Left: Solvation of a polar dendrimer in a protic solvent ("good"), solvent leading to extended conformation exposing a polar surface. Right: Polar dendrimer in an apolar aprotic solvent ("poor"), solvent leading to exposure of an apolar surface consisting of alkyl chains by back-folding*

In dendrimers with an interior structure based on chiral mixed pyridine-dicarboxyanilide structures capable of hydrogen bonding, CD measurements showed that the dendrons were more temperature sensitive to unfolding processes in a polar solvent like acetonitrile compared to apolar solvents.[65] This may be explained from their more open and flexible structure, more easily accessible to solvation and H-bond disruption by polar solvents. The higher generation (G3) dendrons formed a more stable intramolecular network less prone to be "denaturated" by the solvent, resulting in higher denaturation temperatures for these dendrons.

When taking a look at dendrimers with less polar interior structures, *e.g.* dendrimers based on Fréchet type dendrons, the behaviour in various solvents is, as would be expected, significantly different from the more polar dendrimer constructs. For these, rather apolar π-reactive dendrimers, toluene proved to be a "good" solvent because of its ability to solvate the benzene containing Frechet dendrons by π-interactions. In toluene, the hydrodynamical volume was increased from G1 to G4 with strongest effect observed for the lower generations.[66] The increased solvation of the lower generations compared to higher generations may be a consequence of the more open structure of the low generation dendrimers allowing solvent molecules to penetrate into the interior of the dendrimer. A more polar solvent like acetonitrile, with a poor capability to solvate the dendrons, leads to a decrease in hydrodynamical volume indicative of increased intramolecular π–π interactions. The decrease in hydrodynamical volume was most pronounced for the G4 dendrimers.

Dendrimers and the effect of salt: Molecular simulations generally conclude that high ionic strength (high concentration of salts) has a strong effect on charged PPI dendrimers and favours a contracted conformation of dendrimers, with a high degree of back-folding somewhat similar to what is observed upon increasing pH or poor solvation.[67,68] At low salt conditions, the repulsive forces between the charged dendrimer segments results in an extended conformation in order to minimise charge repulsion in the structure (Figure 1.17).

Dendrimers and the effect of concentration: In dendrimers with flexible structures the conformation is not only affected by small molecules like solvents, salts or protons, but may also be sensitive to larger objects, such as other dendrimers or surfaces which can

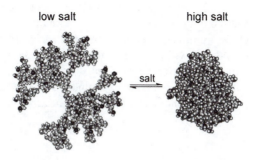

low salt   high salt

salt

Figure 1.17 *Showing the three-dimensional conformational change of a PPI dendrimer upon increasing ionic strength*
(Reprinted from *Chemical Reviews*, 1999, **99**, 1665–1688, with permission. ©1999, American Chemical Society)

have a great affect on the molecular density and conformation of the dendrimer. Small angle X-ray scattering (SAXS) experiments performed on PPI dendrimers (G4, G5) in a polar solvent like methanol show that the molecular conformation of dendrimers upon increasing concentration becomes increasingly contracted. This molecular contraction may minimise the repulsive forces between the dendrimer molecules and increase the ability of the dendrimers to exhibit a more tight intermolecular packing.[69]

1.6 Summary

Dendrimers pose an exciting possibility for chemists to create macromolecular structures with a specifically tailored function or several functions. Dendrimers, like macromolecules found in biology, respond to the surrounding chemical environment showing altered conformational behaviour upon changes in, *e.g.* pH, solvent polarity and ionic strength.

When going from smaller dendritic structures to more globular macromolecular structures, compartments arise and the core region becomes increasingly shielded off from the surroundings by the dendritic wedges and an increasingly dense surface. The built-up dendrimer may be tailored to create a densely packed "De Gennes shell", *e.g.* by the introduction of hydrogen bonding surface groups or a more loose, flexible structure can be obtained by diminishing the attractive forces between the surface functionalities. In flexible dendrimer structures, back-folding may occur as a consequence of weak forces between the surface functionalities or dendrons leading to a more disordered conformation favoured by entropy, where the molecular density is spread out over the entire molecular area. However, back-folding may also be a result of attractive forces (ion-pairing, hydrogen bonding, π-interactions, *etc.*) between functional groups at the inner part of the dendrons and the surface functional groups. In these cases, back-folding is to a large extend driven by enthalpy. However, in both cases, the back-folded state may lead to a more low-energy state of the dendrimer. In addition, the degree of back-folding is to a large extend determined by the surroundings (solvent polarity, ionic strength), thereby constituting a delicate balance between intramolecular forces and forces applied by the surroundings.

The microenvironment in the core may be used to carry low-molecular substances, *e.g.* drugs, or may be useful to create altered properties of core-chromophores or fluorophores, *etc.* Furthermore, the dendrimers expose a multivalent surface, which as elsewhere in biology, is a promising motif to enhance a given functionality. In the next section the multivalency will be treated in more detail, how does the multivalency of the surface functionalities affect a given surface function in biological systems and how does these highly synthetic macromolecules interact with biological systems like cells, proteins and biological membranes *in vitro* and *in vivo*?

References

1. W.F. Ganong, *Review of Medical Physiology*, 15th edn, Prentice-Hall, New York, 1991.
2. K. Autumn, Y.A. Liang, S.T. Hsieh, W. Chan, T.W. Kenny, R. Fearing and R.J. Full, *Nature*, 2000, **405**, 681–685.

3. E. Buhleier, W. Wehner and F. Vögtle, *Synthesis*, 1978, 155–158.
4. D.A. Tomalia, J.R. Dewald, M.R. Hall, S.J. Martin and P.B. Smith, *Preprints 1st SPSJ Polym. Conf.*, *Soc. Polym. Sci. Jpn.*, *Kyoto*, 1984, 65.
5. D.A. Tomalia, H. Baker, J. Dewald, M. Hall, G. Kallos, S. Martin, J. Roeck, J. Ryder and P. Smith, *Polym. J.*, 1985, **17**, 117–132.
6. C. Wörner and R. Mühlhaupt, *Angew. Chem.*, 1993, **105**, 1367.
7. E.M.M. De Brander van den Berg and E.W. Meijer, *Angew. Chem.*, 1993, **105**, 1370.
8. J.P. Tam, *Proc. Natl. Acad. Sci. USA*, 1988, **85**, 5409.
9. F. Zeng and S.C. Zimmernan, *Chem. Rev.*, 1997, **97**, 1681.
10. A.W. Bosman, H.M. Janssen and E.W. Meijer, *Chem. Rev.*, 1999, **99**, 1665.
11. G.M. Dykes, *J. Chem. Technol. Biotechnol.*, 2001, **76**, 903.
12. K. Sadler and J.P. Tam, *Rev. Mol. Biotechnol.*, 2002, **90**, 195.
13. M.J. Cloninger, *Curr. Opin. Chem. Biol.*, 2002, **6**, 742.
14. U. Boas and P.M.H. Heegaard, *Chem. Soc. Rev.*, 2004, **33**, 43.
15. T. Emrick and J.M.J. Fréchet, *Curr. Opin. Colloid. Interface Sci.*, 1999, **4**, 15.
16. D.A. Tomalia, D.M. Hedstrand and M.S. Ferritto, *Macromolecules*, 1991, **24**, 1435.
17. C. Hawker and J.M.J. Fréchet, *J. Chem. Soc. Chem. Commun.*, 1990, 1010.
18. C.J. Hawker and J.M.J. Fréchet, *J. Am. Chem. Soc.*, 1990, **112**, 7638.
19. C.J. Hawker, K.L. Wooley and J.M.J. Fréchet, *J. Chem. Soc. Perkin. Trans. 1*, 1993, 1287.
20. J.P. Tam, Synthesis of peptides and peptidomimetics, Houben-Weyl Methods of organic chemistry, in *Peptide Dendrimers and Protein Mimetics*, M. Goodman (ed), Thieme, Stuttgart, 2000.
21. Commercially available by dendritech, www.dendritech.com.
22. G.R. Newkome, C.N. Moorefield, G.R. Baker, M.J. Saunders and S.H. Grossman, *Angew. Chem. Int. Ed. Eng.*, 1991, **30**, 1178.
23. J.-P. Majoral and A.-M. Caminade, *Chem. Rev.*, 1999, **99**, 845.
24. W.B. Turnbull and J.F. Stoddart, *Rev. Mol. Biotechnol.*, 2002, **90**, 231.
25. R.H.E. Hudson and M.J. Damha, *J. Am. Chem. Soc.*, 1993, **115**, 2119.
26. R.H.E. Hudson, S. Robidoux and M.J. Damha, *Tetrahedron Lett.*, 1998, **39**, 1299.
27. T.W. Nilsen, J. Grazel and W. Prensky, *J. Theor. Biol.*, 1997, **187**, 273.
28. G.R. Newkome, E. He and C. Moorefield, *Chem. Rev.*, 1999, **99**, 1689.
29. M.K. Lothian-Tomalia, D.M. Hedstrand and D.A. Tomalia, *Tetrahedron*, 1997, **53**, 15495.
30. J.R. Morgan and M.J Cloninger, *Curr. Opin. Drug Discov. Develop.*, 2002, **5**, 966.
31. F. Vögtle, H. Fakhrnabavi, O. Lukin, S. Müller, J. Friedhofen and C.A. Schally, *Eur. J. Org. Chem.*, 2004, 4717.
32. E.T. Kaiser, H. Mihara, G.A. Laforet, J.W. Kelly, L. Walters, M.A. Findeis and T. Sasaki, *Science*, 1989, **243**, 187.
33. K.L. Wooley, C.J. Hawker and J.M.J. Fréchet, *J. Am. Chem. Soc.*, 1993, **115**, 11496.
34. S.C. Zimmerman, F. Zeng, D.E.C. Reichert and S.V. Kolotuchin, *Science*, 1996, **271**, 1095.

35. A.W. Freeman, R. Vreekamp and J.M.J. Fréchet, *Abstr. Pap. Am. Chem. Soc.*, 1997, **214**, 128-PMSE.

36. P.S. Corbin, L.J. Lawless, Z. Li, Y. Ma, M.J. Witmer and S.C. Zimmerman, *Proc. Natl. Acad. Sci. USA*, 2002, **99**, 5099.

37. C.R. DeMattei, B. Huang and D.A. Tomalia, *Nano Lett.*, 2004, **4**, 771.

38. Y. Choi, T. Thomas, A. Kotlyar, M.T. Islam and J.R. Baker, *Chem. Biol.*, 2005, **12**, 35.

39. M. Kawa and J.M.J. Fréchet, *Chem. Mater.*, 1998, **10**, 286.

40. V.V. Narayanan and E.C. Wiener, *Macromolecules*, 2000, **33**, 3944.

41. V. Percec, C.-H. Ahn, G. Ungar, D.J.P. Yeardley, M. Möller and S.S. Sheiko, *Nature*, 1998, **391**, 161.

42. M. Numata, A. Ikeda and S. Shinkai, *Chem. Lett.*, 2000, 370.

43. T.E. Creighton, *Proteins, Structures and Molecular Properties*, 2nd edn, W.H. Freeman, New York.

44. C. Kojima, K. Kono, K. Maruyama and T. Takagishi, *Bioconjugate Chem.*, 2000, **11**, 910.

45. D. Farin and D. Avnir, *Angew. Chem. Int. Ed. Engl.*, 1991, **30**, 1379.

46. A. Nourse, D.B. Millar and A.P. Minton, *Biopolymers*, 2000, **53**, 316.

47. G.M. Pavlov, E.V. Korneeva and E.W. Meijer, *Colloid Polym. Sci.*, 2002, **280**, 416.

48. V. Percec, W.-D. Cho, P.E. Mosier, G. Ungar and D.J.P. Yeardley, *J. Am. Chem. Soc.*, 1998, **120**, 11061.

49. M. Ballauf, *Topics Curr. Chem. Dendrimers III: Design, dimension, function.* 2001, **212**, 177.

50. M. Ballauf and C.L. Likos, *Angew. Chem. Int. Ed. Engl.*, 2004, **43**, 2998.

51. P.G. DeGennes and H. Hervet, *J. Phys. Lett. Paris*, 1983, **44**, L351.

52. J.F.G.A. Jansen, E.M.M. De Brabander van den Berg and E.W. Meijer, *Science*, 1994, **266**, 1226.

53. A.W. Bosman, M.J. Bruining, H. Kooijman, A.L. Spek, R.A.J. Janssen and E.W. Meijer, *J. Am. Chem. Soc.*, 1998, **120**, 8547.

54. R.L. Lescanet and M. Muthukumar, *Macromolecules*, 1990, **23**, 2280.

55. A.D. Meltzer, D. Tirrel, A.A. Jones, P.T. Inglefield, D.M. Hedstrand and D.A. Tomalia, *Macromolecules*, 1992, **25**, 4541.

56. A.D. Meltzer, D. Tirrel, A.A. Jones and P.T. Inglefield, *Macromolecules*, 1992, **25**, 4549.

57. M. Murat and G.S. Grest, *Macromolecules*, 1996, **29**, 1278.

58. M. Elshakre, A.S. Atallah, S. Santos and S. Grigoras, *Comput. Theor. Polym. Sci.*, 2000, **10**, 21.

59. I. Lee, B.D. Athey, A.W. Wetzel, W. Meixner and J.R. Baker, *Macromolecules*, 2002, **35**, 4510.

60. T. Terao and T. Nakayama, *Macromolecules*, 2004, **37**, 4686.

61. D.J. Wang and T. Imae, *J. Am. Chem. Soc.*, 2004, **126**, 13204.

62. I.B. Rietveld, W.G. Bouwman, M.W.P.L. Baars and R.K. Heenan, *Macromolecules*, 2001, **34**, 8380.

63. M. Chai, Y. Niu, W.J. Youngs and P.L. Rinaldi, *J. Am. Chem. Soc.*, 2001, **123**, 4670.

64. J.E. Butler, L. Ni, R. Nessler, K.S. Joshi, M. Suter, B. Rosenberg, J. Chang, W.R. Brown and L.A. Cantaro, *J. Immunol. Methods*, 1992, **150**, 77.
65. J. Recker, D.J. Tomcik and J.R. Parquette, *J. Am. Chem. Soc.*, 2000, **122**, 10298.
66. S. De Backer, Y. Prinzie, W. Verheijen, M. Smet, K. Desmedt, W. Dehaen and F.C. De Schryver, *J. Phys. Chem. A*, 1998, **102**, 5451.
67. P. Welch and M. Muthukumar, *Macromolecules*, 1998, **31**, 5892.
68. A. Ramzi, R. Scherrenberg, J. Joosten, P. Lemstra and K. Mortensen, *Macromolecules*, 2002, **35**, 827.
69. A. Topp, B.J. Bauer, T.J. Prosa, R. Scherrenberg and E.J. Amis, *Macromolecules*, 1999, **32**, 8923.

CHAPTER 2

Properties of Dendrimers in Biological Systems

2.1 Significance of Multivalent Binding in Biological Interactions

Nature utilises dendritic structures in a wide range of applications in the design of animals and plants. The reason is the beneficial behaviour of these dendritic structures with their ability to expose a highly multivalent surface to ensure maximum interaction with the surroundings.

Multivalent interaction between substrates and various receptors is a common theme in biology, where antibodies divalently bind specific antigens, and viruses adhere to target cells prior to infection via multivalent interactions between viral trimeric haemagglutinin and multiple sialic acid carbohydrate residues on the surface of the target cell.[1]

The favourable binding obtained in multivalent systems may be explained from basic thermodynamics, stating that the favourable course of a reaction leads to an increase in entropy ($\Delta S > 0$).[2] However, the loss of entropy upon binding of a polyvalent ligand to a receptor is greatly diminished because the degrees of motional freedom of the ligands "interacting sites" are significantly reduced because of the anchoring effect (chelate effect) obtained by binding the first "arm" of the ligand to the receptor (Figure 2.1). The decreased entropy loss of binding the second ligand arm makes this a highly favourable reaction from a thermodynamic point of view, thus shifting the equilibrium strongly towards binding of this second ligand site.

The phenomenon of binding-affinities being much larger in multivalent interactions than the expected additive affinity increase based on the system valency, has been given different terms depending on the scientific context encountering and treating this effect. In inorganic coordination chemistry, the strong interaction between multivalent (or multidentate) ligands with metal ions is denoted "the chelate effect" where the most widely used ligand based on this property may be the very well-known hexavalent ligand "EDTA" used for vast amount of applications due to its strong complexation to divalent earth metals.

In glycobiology, the multivalent interaction between numerous carbohydrates entities on a cell surface with numerous lectin molecules on *e.g.* a viral surface has

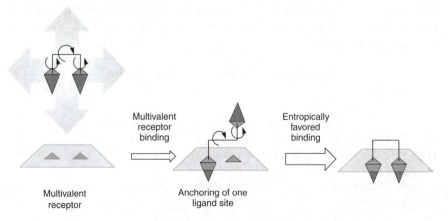

Figure 2.1 *The binding of a divalent ligand to a divalent receptor. The arrows show some of the rotational and some translational degrees of freedom (shaded arrows in the back). By the binding of the first ligand site, the degrees of freedom are significantly reduced, leading to an entropically favoured binding of the second ligand site*

been termed "the glycoside cluster effect". And finally, in dendrimer chemistry the synergistic enhancement of binding as a consequence of multivalency is termed "the dendritic effect". Roughly spoken, these three terms all treat the same subject, albeit, in systems of different molecular size and properties. The term "chelate effect" is usually applied for smaller inorganic or organic systems having sub-nanometer dimensions. The "dendritic effect" is applied to larger dendritic molecular systems usually in nanometer dimensions, and finally the "cluster glycoside" effect is a term usually applied on supramolecular chemical (and biological) events taking place in systems having nano- to micrometer dimensions.

In biology, multivalent interactions are successfully applied where it is difficult to obtain high-affinity binding to a monovalent ligand. In these cases, the synergistic increase in affinity of a polyvalent ligand/receptor results in an overall significant affinity.

2.1.1 Dendritic Effect

A dendritic- or cluster effect is apparent when a simultaneous binding of entities of a ligand leads to a synergistic increase in the overall association as result of a cooperative binding of ligand sites,[1] so that

$$K_{\mathrm{poly}} \, (\mathrm{synergy}) > (K_{\mathrm{mono}})^n$$

where n equals number of ligand sites.

In contrast, the additive increase of binding affinity observed with a polyvalent ligand–receptor systems where no synergy (cooperative) effect is observed, equals:

$$K_{\mathrm{poly}} \, (\mathrm{additive}) = (K_{\mathrm{mono}})^n$$

Polyvalent interactions of high-affinity ligands lead to a higher specificity in comparison to ligands with lower affinity because of the resulting in an exponential increase of binding affinity

$$K_{mono} \text{ (lower affinity)} < K_{poly} \text{(lower affinity)}$$

$$K_{mono} \text{(higher affinity)} \ll K_{poly} \text{ (higher affinity)}$$

High specificity is crucially important in the complex molecular recognition found in biology.[3]

Thus, in nature the multivalent binding/presentation of a certain motif is widely used to provide:

- tight binding of ligands, which in their monovalent form only bind to the receptor with low affinity and
- more efficient cell–cell, cell–virus or cell–bacteria interactions.

2.1.2   Carbohydrate Ligands

Dendrimers provide a simple synthetic scaffold for multivalent presentation of a certain motif together with an easy access to modulate the spatial arrangement of each ligand site in a well-defined way, and several research groups have investigated the enhancement of binding between ligands based on dendrimer scaffolds with various receptors.

Especially, in the weak interactions between carbohydrates and proteins (*e.g.* lectins or antibodies), the ability to create a strong carbohydrate–protein affinity may be provided by increasing the valency of the carbohydrate-containing ligand, *i.e.* by coupling the monomeric carbohydrate ligand to the dendrimer surface for multivalent presentation. The dendrimer scaffold offers an easy access to vary the spatial arrangement of the carbohydrate ligands to make a perfect fit for a particular receptor. From a medicinal point of view, the interactions between carbohydrates and proteins (lectins) is of high importance, as many biological events involve specific protein/carbohydrate recognitions *e.g.* in virus/cell and bacteria/cell interactions. In addition to the understanding and careful control of these recognition processes, dendrimers may pose a very important platform in the development of drugs and vaccines, which may serve as inhibitors for microbial invasion or as recognition targets for the generation of antibodies.

The well-known lectin Concanavalin A (Con A), which recognises gluco- and mannopyranoside moieties present in carbohydrate-containing ligands, has been subjected to numerous investigations on the effect of binding multimerically presented carbohydrates in comparison to binding the monovalent ligand.[4–7] The binding between this lectin and glucose/mannose containing substrates shows tremendous increase (600-fold) in the binding to Con A upon going from monovalent mannopyranosides to a 16-mer presentation on polyglycine/lysine-based dendrimers.[5] Interestingly, it was found in addition that dendrimers having the

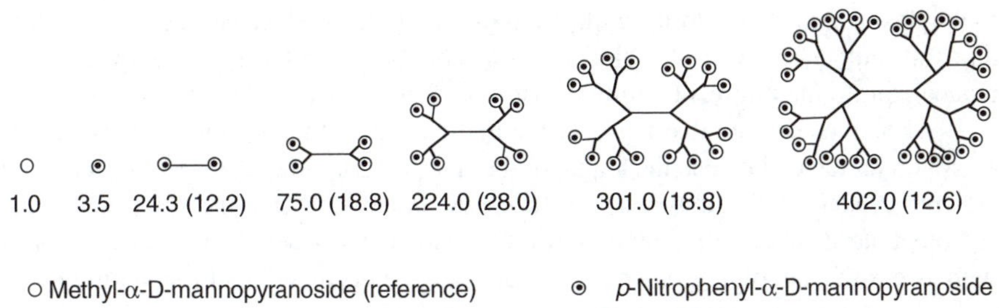

O	⊚	⊚—⊚				
1.0	3.5	24.3 (12.2)	75.0 (18.8)	224.0 (28.0)	301.0 (18.8)	402.0 (12.6)

O Methyl-α-D-mannopyranoside (reference)   ⊚   *p*-Nitrophenyl-α-D-mannopyranoside

Figure 2.2 *Schematic depiction on the increase in binding-affinity between mannosylated dendrimer and Con A with increasing valency of mannosylated polyglycine-lysine dendrimers in comparison to monomeric methyl-a-D-mannopyranoside. The numbers in brackets refer to the increase in binding-affinity per mannose unit*

carbohydrate bound via an aromatic spacer were bound significantly better to the lectin in comparison to a similar dendrimer carrying its carbohydrates via an aliphatic spacer. This indicates that even small modulations in the molecular motif have an effect on the recognition by the lectin.

Where the previously mentioned polyglycine–lysine-based dendrimers had to be synthesised step by step using divergent solid-phase synthesis, preformed PAMAM dendrimers could constitute a useful scaffold for the subsequent surface modification with a lectin substrate like mannose. In comparison to the polyglycine/lysine dendrimers, the mannosyl-coated PAMAM dendrimers showed somewhat lower affinities, albeit, the binding-affinity towards Con A was increased 400 times when applying the mannosylated G4-PAMAM dendrimer (32-mer) compared to monomeric mannose (Figure 2.2).[6]

Also, the binding of carbohydrate-containing substrates to another biologically important lectin, the Shiga-like toxin (SLT), has been shown to be dependent on the valency of the substrate. SLT, or verotoxin is an exotoxin of *Escherichia coli*, resembling the exotoxin of *Shigella dysenteriae*, and responsible for haemolytic uremic syndrome (HUS) upon infection with certain serotypes of *E. coli* expressing SLT leading to renal dysfunction. A panel of monovalent and divalent carbohydrate substrates showed preferable binding of the divalent substrates.[8] Recently, further investigation on the multivalent binding between the SLT system and silicon-containing dendrimers with a highly multivalent surface has shown further enhancement in the binding of this toxin. These conjugates are suggested to have a therapeutic effect in the treatment of the infection caused by these bacteria. The use of carbohydrate-exposing dendrimers (glycodendrimers) as antibacterial drugs and therapeutics will be treated more thoroughly in Chapter 4.

Adhesins are a class of carbohydrate-binding proteins located on the fimbriae of certain bacteria, and are important tools for the initial recognition/adhesion of bacteria to the cellular surfaces of the host organism. The use of multivalency in high-affinity ligands for bacterial adhesins (antiadhesins) may be an important platform in the design of antibacterial therapeutics for prevention of infection and for minimising the pathological symptoms inflicted upon the organism by the bacteria (Figure 2.3).

Polylysine-based dendrons with mannosylated surfaces, have shown a 12.500 times affinity increase in the binding to *E. coli* type 1 adhesin (FimH protein) in comparison to the reference inhibitor Me-α-D-mannopyranoside.[9]

These studies showed that not only the multivalency of the synthetic inhibitor was an important factor, but also the length of the spacer separating the respective mannose moieties. Also here, the presence of an aromatic- or aliphatic spacer moiety next to the carbohydrate increased the binding affinity significantly, whereas more polar groups such as positive amidino group or an amide lead to a decrease in adhesion binding.

As we will see in Chapter 4, the use of carbohydrates for surface modification of dendrimers as well as smaller negatively and positively charged functional groups is widely used, not only in the competitive binding of bacteria and viruses to hamper their adhesion to the host cell, but also in destruction of the microbial intruder.

2.2   Biocompatibility of Dendrimers

In order to incorparate dendrimers as biological agents, introducing them to biological systems, certain properties have to be present or fine-tuned through preclinical chemical modification. The demands for dendrimers to obtain an appropriate biological function are that they should be:

- non-toxic;
- non-immunogenic (if not required for vaccine purposes);
- biopermeable, having the ability (if required) to cross biobarriers, *e.g.* the intestines, blood-tissue barriers, cell membranes or bacterial membranes *etc.*;
- able to stay in circulation in the biological system for the time needed to have the desired clinical effect;
- able to target specific biological structures.

Biological properties like the toxicity or the immunogenicity profile of a dendrimer is to a large extent governed by the size of the dendrimer and by the surface groups present on the particular dendrimer. The inner dendritic structures are generally of less importance as interactions between the dendrimer and the surroundings generally take place via the groups exposed on the dendrimer surface, which may enable the dendrimer to penetrate *e.g.* a cell surface in a functional or disruptive way (*i.e.* to induce endocytosis or cytotoxicity).

Dendrimers are in this regard an interesting class of molecules, because they relatively easily may be modified at their surfaces to obtain specific biological properties. Therefore, by appropriate variation of the surface motifs, dendrimers with specific biological properties can be constructed and their biological properties can be modulated in endless ways (Figure 2.4).

2.3   *In Vitro* Cytotoxicity of Dendrimers

The vast majority of biological studies on dendrimers concern the measurements of *in vitro* cytotoxicity. Performing the studies *in vitro* gives a direct measure of the effects of a specific molecular motif or change in motif, without the complexity

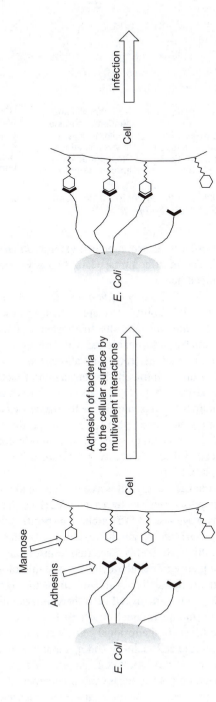

Figure 2.3 *Schematic depiction of a bacterial infection by E. Coli, initiated by multivalent interactions between adhesins on the bacterial fimbriae and multiple mannosyl residues on the cellular surface*

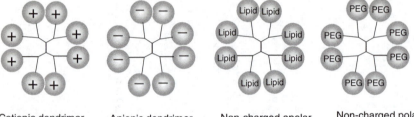

Cationic dendrimer surface	Anionic dendrimer surface	Non-charged apolar dendrimer surface	Non-charged polar dendrimer surface
In vitro toxicity: +	In vitro toxicity: -	In vitro toxicity: +	In vitro toxicity: -
In vivo tox effects: +	In vivo tox effects: -	In vivo tox effects: -	In vivo tox effects: -
Biopermeability: +	Biopermeability: -/+	Biopermeability: +	Biopermeability: -
immunogenic: +	immunogenic: -	Immunogenic: +	Immunogenic: -

Figure 2.4 *Dendrimers "masked" with different surface groups, leading to significant modulation of their biological properties*

often experienced *in vivo*. In order to measure the cell viability *in vitro*, a number of analytical methods are used. The most common viability assays, and terms in this regard, will briefly be mentioned here.

LDH assay: Together with the MTT assay (*vide infra*), this is the most frequently utilised assay for measuring cell viability. Lactate dehydrogenase (LDH) is an enzyme present in the cytosol of mammalian cells. In intact cells, the cell membrane is not permeable for LDH, which resides inside the cell. However, upon cell damage, the breakdown of the cell membrane (cell-lysis) allows the LDH to diffuse out into the surroundings. LDH catalyses intracellular conversion of lactate to pyruvate under the formation of NADH from NAD^+. The spectroscopic changes (340 nm) resulting from this reaction may be used as a measure for the cell viability.[10] Further development of this assay relies on the fact that the NADH produced by this reaction have the ability to reduce tetrazolium salts into strongly coloured formazan adducts, and the degree of cell-lysis/cell viability can be measured at *e.g.* 492 nm with high sensitivity (see Figure 2.5.)

MTT assay: Like some of the LDH assays, this assay relies on the reductive cleavage of the tetrazolium moiety to formazan by LDH/NADH or the mitochondrial enzyme SDH (succinate dehydrogenase). MTT refers to a specific tetrazolium compound (3-[4,5-methylthiazol-2-yl]-2,5-diphenyltetrazolium bromide). In this assay the MTT is taken into the healthy cell, and is converted to insoluble formazan crystals inside the cell as a consequence of cellular cytosolic and mitochondrial activity. In an intact cell, the formazan precipitate is not able to diffuse through the cell membrane and resides inside. However, in damaged cells the formazan diffuses out and the intracellular content of formazan is much lower in this case. The content of viable cells is analysed by cell-lysis of the residual cells, where the amount of residual formazan is proportional to the cell viability and metabolic activity.[11]

Haemolysis assay: In this assay, red blood cells are exposed to the target compound, and the release of haemoglobin is measured in supernatant (550 nm). The degree of haemolysis is expressed as the percentage of released haemoglobin induced by 1% v/v Triton-X-100.

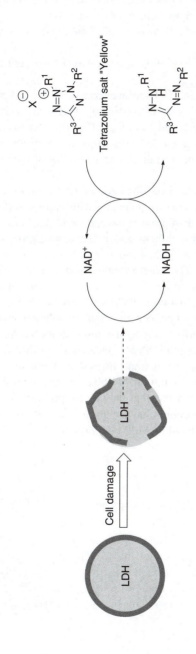

Figure 2.5 *LDH assay, measuring the cell viability*

Scanning electron microscopy (SEM) is an alternative technique for the visualisation of the cell viability. Upon imposing a cytotoxic substance to a cell –culture, an increased amount of "clothing" of the cells can be observed by this technique, which indicates poorer cell viability and prelytic behaviour of the cells.

IC_{50}: The concentration (in $\mu g \ mL^{-1}$) causing 50% inhibition of *e.g.* enzyme activity.

LD_{50}: The concentration lethal for 50% of the test animals (*in vivo* studies) or cells in a cell culture (*in vitro*).

Surface groups inducing cytotoxicity: Cell surfaces in eukaryotic and prokaryotic organisms expose negatively charged surfaces consisting of *e.g.* sialic acid carbohydrate moieties. In order to avoid disintergration of the cell surface due to electrostatic repulsion between the negatively charged surface groups, the presence of divalent earth metal cations *e.g.* Ca^{++} ensures overall charge neutrality of the cell surface (Figure 2.6).

Dendrimers or other macromolecules exposing cationic (*e.g.* basic) surfaces generally show cytotoxic activity. The toxicity of molecules with cationic surface groups (*e.g.* primary amines) is attributed to disruption of the cell membrane through the initial adhesion by electrostatic attraction to the negative cell surface groups, followed by either hole formation or endocytosis.[12] The formation of holes and channels in the cell wall will cause the cell to lyse. The exact mechanism of the destabilisation of the cell wall caused by cationic molecules has, however, not been elucidated in detail. Model studies of interaction between PAMAM dendrimers and liposome vesicles show disruptive interactions between the lipid membrane and the dendrimers and indicate that large dendrimer aggregates are responsible for the membrane permeation. However, it is suggested that transfection and disruption of the cell membranes happens via endosome formation, which may enable the dendrimers to disrupt the cell membrane from inside the cell.[13] The cytotoxicity profile of these cationic dendrimers seems to be governed extensively by the primary amine surface groups as *e.g.* melamine-based dendrimers having amine surface groups have an *in vitro* toxicity similar to amino PPI and PAMAM dendrimers.[14]

Outside

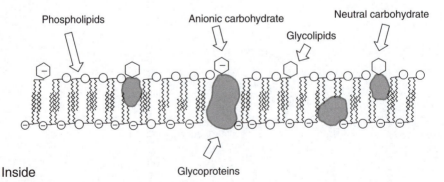

Inside

Figure 2.6 *Eukaryotic cell wall*

Unmodified amino-terminated G4-PAMAM dendrimers were shown to be more toxic towards muscle cells, *i.e.* to show higher myotoxicity compared to cationic liposomes and proteins. Also, neuroblastoma cells suffered from cell-lysis after 1 week of exposure to amino-terminated PAMAM and PPI dendrimers.[15]

The ability for the dendrimer to expose a cationic surface is proportional to its cytotoxicity. It has been shown that cytotoxicity of dendrimers having amine surface groups is highly dependent on the degree of substitution on the surface amine, *i.e.* primary amines are more toxic than secondary and tertiary alkyl amines. In the more substituted amines, the positive charge on the nitrogen atom is to a larger extent shielded off from interacting with the surroundings by the alkyl substituents, which have a larger size compared to the hydrogen atoms present on a primary amine.[16,17]

Dendrimers having cationic guanidinium surface groups are generally found to be cytotoxic and haematotoxic even at low concentrations,[18] however, dendrons based on an aliphatic skeleton carrying guanidinium surface groups show no significant toxicity, although in this specific study, only a G2 dendron was tested. On the other hand, these dendrons were shown to be good candidates for delivery through intracellular translocation.[19]

Whereas cationic amino-terminated dendrimers show significant cytotoxicity, also apolar functionalities presented (*e.g.* aromates or lipid chains) on a dendrimer surface may induce cell-lysis.[20] The mechanism of cell membrane disruption may in these cases be governed by weaker hydrophobic interactions between the lipid bilayer in the cell membrane and the dendrimer surface lipids. However, as we shall see, lipophilic surfaces on dendrimers may also play a favourable role in vesicle and endosome formation, *i.e.* in membrane penetration by endocytosis.

Surface groups decreasing cytotoxicity: Encapsulation or "quenching" of the charged surface amines by *e.g.* the previously mentioned alkylation or amidation strongly decrease the cytotoxcicity of the dendrimer as a consequence of the alkyl groups' ability to shield off the basic nitrogen atoms. Also, for linear polymers like poly-L-lysine it has been found that the toxicity may be quenched by Michael addition to *N,N*-dimethylacrylamides, creating a non-toxic alkyl-amide surface.[21]

Investigations on dendrimers having a trialkylbenzene core show that these types of dendrimers are cytotoxic (MCF-7 human breast cancer cells) when covered with phenylalanine surface groups, probably due to disruption of the cell membranes via electrostatic and/or hydrophobic interactions provided by the basic primary amines and the phenyl groups present at the dendrimer surface (Figure 2.7). Upon introduction of dansyl surface groups on these dendrimers, the cytotoxicity was quenched, even though the dansyl group, like the phenylalanine, contains both an aromatic and a basic part. However, the dansyl group contains an aromatic tertiary amine, where the basic nitrogen is shielded off by the substituents, and in addition has a lower basicity compared to phenylalanine. These factors lead to a significantly less-charged surface, although this surface still can interact with the cell membrane through hydrophobic interactions.[22] Dendrimers partially derivatised with dansyl groups, thus having some primary amines present at their surface still maintained significant cytotoxicity. Surprisingly, the authors found that surface derivatisation of these dendrimers with an ethylene diamine derivative leads to low cytotoxicity as well. This finding contradicts the general assumption that dendrimers carrying

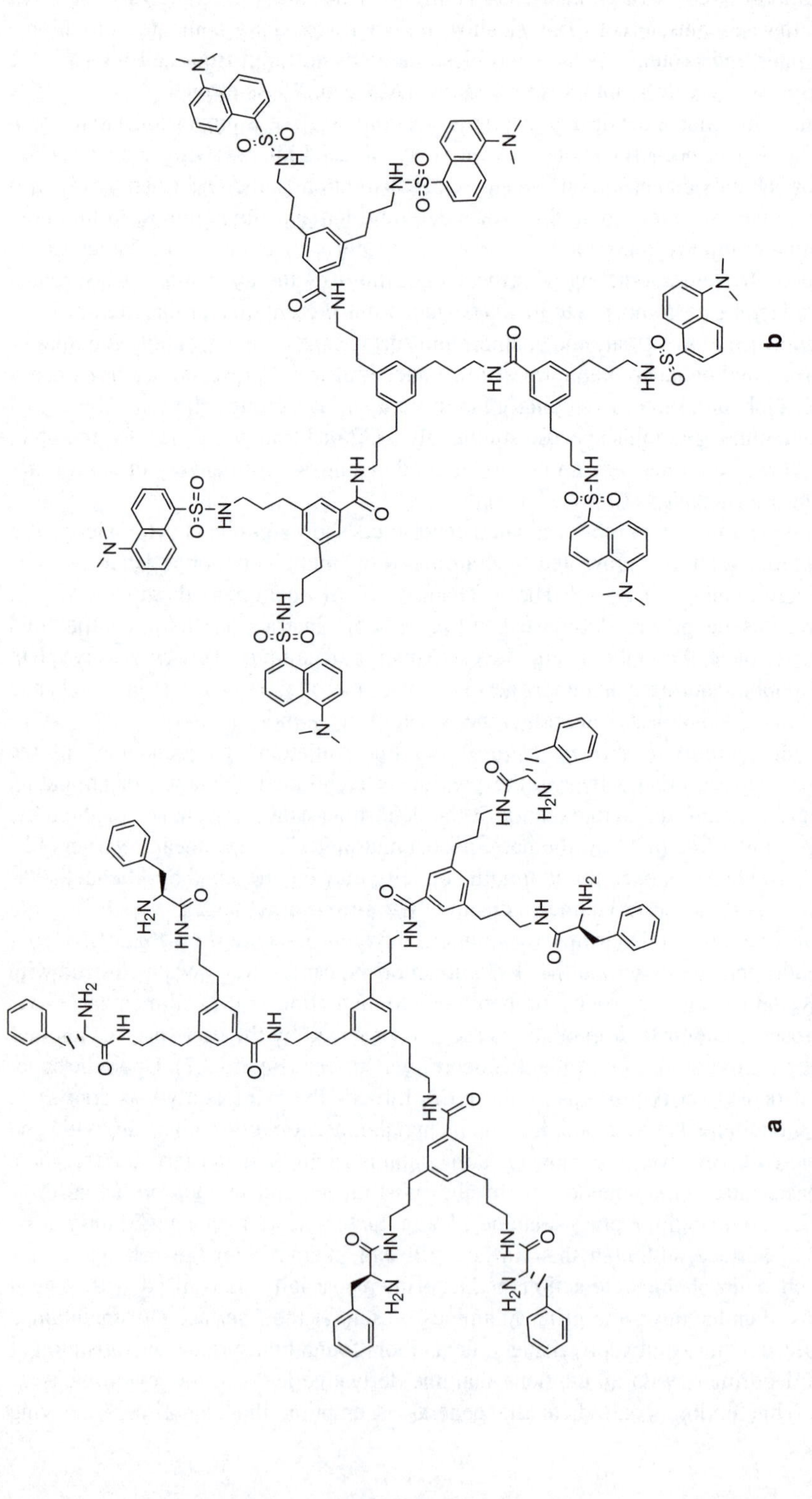

Figure 2.7 *Two trialkylbenzene-based dendrimers having surfaces consisting of an aromatic and an amino residue. However, whereas dendrimer **a** is highly cytotoxic, dendrimer **b** only has low cytotoxicity against MCF-7 human breast cancer cell-line[22]*

primary amines at their surfaces are cytotoxic. The explanation for the behaviour of these diamine-derivatised dendrimers is not straightforward, but may rely on the higher ability of quenching the 1.2 diamines membrane binding because of the binding to metal ions or hydrogen donors as *e.g.* water (chelate effect).[22]

Converting the basic primary amines to non-charged amides by acylation and acetylation strongly affects the toxicity profile towards low toxicity and poor adhesion to the cell membrane.[12] Even partial derivatisation of the surface primary amines to amides carrying a diminished positive charge results in decreased toxicity of the dendrimer (Figure 2.8).

The cytotoxicity profile of PAMAM dendrimers may be fine-tuned towards low toxicity (from IC50 ~ 0.13 to >1 mM, Caco-2 epithelial cells) by mixed derivatisation with lipid and PEG chains. Increasing the number of lipid chains results in increased cytotoxicity, although, the lipidated dendrimers have lower cytotoxicity in comparison with the unmodified amino-terminated dendrimers.[20]

For dendrimers having a triazine-based scaffold, the cytotoxicity was found roughly to decrease by the following sequence of surface groups (with their respective charges under physiological conditions).[18]

$$NH_3^+ > Guanidyl^+ > SO_3^- > PO_3^{2-} > COO^- > PEG$$

When covering an *e.g.* PAMAM or PPI dendrimer surface with anionic groups like carboxylates or sulfonates, dendrimers with very low or no cytotoxicity are obtained. The general non-toxic nature of anionic functionalities may be due to the non-adhesive nature of these dendrimers towards the cellular membranes being negatively charged as well. Also, haemolysis is greatly diminished with the introduction of anionic groups at the dendrimer surface. As we shall see, these anionic surface groups may even have a therapeutic effect against *e.g.* viral or bacterial infections.

Hydroxy groups or hydroxyl-containing compounds *e.g.* carbohydrates at the dendrimer surface generally result in low cytotoxicity of the dendrimers regardless of the dendrimer scaffold. PAMAM dendrimers with hydroxyl surface groups (PAMAM–OH) have in some cellular test systems even been found to exhibit a lower cytotoxicity compared to carboxy-terminated PAMAM dendrimers of similar generation.[23]

Similarly, PAMAM dendrimers surface modified with fructose and galactose show good stabilisation properties for hepatocytes (liver cells) *in vitro*. The analysed liver cells had better viability when cultured in the presence of the fructose/galactose-modified dendrimers, whereas pronounced apoptosis and necrosis were experienced by hepatocytes cultured on unmodified amino-terminated PAMAM dendrimers.[24] Also, haematotoxicity studies performed on silicon-based dendrimers having hydroxy surface groups conclude that these dendrimers showed a low haematotoxicity and cytotoxicity, similar to that of carboxy-terminated PAMAM dendrimers.[25]

Introduction of uncharged PEG chains on the dendrimer surface generally leads to very low cytotoxicity indeed. This inertness may be due to the non-disturbing properties of PEG probably caused partly by the ability of the PEG chain to uphold a sufficient hydration of the dendrimer surface together with charge neutrality, thereby suppressing hydrophobic or ionic interactions with the cell membrane.[26]

On the other hand, cytotoxicity studies performed on amino-terminated polylysine (PLL) dendrimers complexed with DNA did not show any clear diminished toxicity of PEGylation of the surface amines, although the authors found that a 4:1 charge ratio of G5-PLL/DNA complex exhibited lower cytotoxicity in comparison to a similar complex comprising of the unmodified dendrimer (20% versus 40%).[27]

Additives decreasing cytotoxicity: Additives may lead to a significant reduction of cytotoxicity when added to toxic dendrimers having cationic surface groups. In an *in vitro* system of human carcinoma (HeLa) cell line it has been shown that formulations of foetal calf serum with an amino-terminated G6-PAMAM dendrimer partially modified with the Oregon Green fluorophore exhibit lower toxicity in comparison to the dendrimer alone.[28] The cytotoxicity of these formulations was reduced further by complexation with oligonucleotides. An explanation for these findings could be the ion pairing between the basic dendrimer and acidic proteins in the serum mixture, giving dendrimer–protein complexes with reduced charge and increased shielding of the dendrimer surface in comparison to the uncomplexed dendrimer. Also, formulations of dendrimers with ovalbumin show lower cytotoxicity compared to free dendrimer.[29] This may be similar to what is observed for serum addition, due to the electrostatic interaction between the negatively charged ovalbumin surface (pI~5) groups and the cationic dendrimer, leading to overall low charged or neutral complexes, similar to what is observed for dendrimer–DNA complexes or dendrimers in presence of foetal calf serum (*vide supra*).

Additives such as DNA or RNA as well as phosphorothioate DNA analogues have been found to greatly reduce the cytotoxicity of dendrimers carrying amino surface groups. This could rely on the reduction of overall positive charge upon complexation to negatively charged DNA compared to the uncomplexed cationic dendrimer.[15,16] Detailed studies of the complexation between dendrimers and DNA have revealed that the overall charge of a dendrimer–DNA complex is not only reduced, but may even be negative in the case of complex formation between a G2-PAMAM dendrimer and DNA. Furthermore, it is shown that the DNA to a large extent is wrapped around the dendrimeric spheres, thereby creating a non-toxic "shield" for the dendrimer cations.[30] The negatively charged DNA may lead to a general shielding of the cationic dendrimer surface amines, similar to what is obtained by covalent modification of the surface amines. However, upon increase of the dendrimer generation the cytotoxicity of the dendrimer/DNA complexes is increased, probably due to greater exposure of positive charges to the surroundings.[31]

Furthermore, these cytotoxicity studies on polycation–DNA complexes showed the same or higher cytotoxicity when an unmodified amino-terminated G5-PAMAM dendrimer was formulated with high concentrations of DNA.[31] In this case, the higher toxicity of the DNA–dendrimer complexes is not directly attributed to toxicity of the cationic amino-terminated dendrimer, but may be related to the cellular stress upon transfection with high levels of DNA ($3 \mu g \, mL^{-1}$), which leads to apoptosis.[32]

Also, for amino-terminated PLL dendrimers (G3, G4) complexation with DNA leads to low cytotoxicity (<20%, MTT assay, retinal pigment epithelial cell line D407). Only the G5-PLL/DNA complexes (charge ratio 4:1) showed significant cytotoxicity (40% decrease in cell viability) in comparison to complexes comprising of lower generation (G <5) dendrimers.[27] As for PAMAM dendrimers, it is rationalised

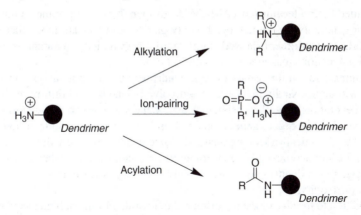

Figure 2.8 *The "detoxification" of primary amines by shielding the positive charge at the nitrogen atom by alkylation reactions (performed e.g. by alkyl halides or Michael addition), ion-pairing with anions (e.g. DNA or proteins) or cation quenching by acylation or similar reactions*

that the higher generation PLL dendrimers due to their more spherical and rigid structure and higher surface charge, still may be able to expose a significant number of positive charges even when complexed to DNA.[31]

These dendrimer–DNA complexes interact and penetrate the cell membrane predominantly by endocytosis, which is a non-toxic membrane penetrating pathway alternative to the cytotoxic "hole formation pathway" accompanied by cell-lysis. The transfection mechanism by endocytosis and endosomal release of the dendrimer–DNA complexes for further delivery to the cell will be more thoroughly explained in Chapter 3.

Effect of the dendrimer scaffold on cytotoxicity: Initial comparative studies between PAMAM (starburst) dendrimers and the polylysine 115 (PLL115) dendrimer performed on different cell lines show that the PAMAM dendrimers had significantly lower cytotoxicity than polylysine 115 (LD50(PLL115) = 25 μg mL^{-1} and LD50(G6-PAMAM) >300 μg mL^{-1}).[33]

Cytotoxicity as well as haematotoxicity of amino-terminated PAMAM and PPI dendrimers has been shown by several research groups to be strongly dependent of the dendrimer generation (*i.e.* the size of the dendrimer scaffold), with the higher generation dendrimers being the most cytotoxic.[25,34] Both dendrimers having PAMAM, PEI and PPI structures show the same trend in generation-dependent cytotoxicity. Zinselmeyer and co-workers[35] found a roughly generation-dependent cytotoxicity profile for PPI dendrimers with the following order:

G5-PPI > G4-PPI > G3-PPI > DOTAP > G1-PPI > G2-PPI

where DOTAP is *N*-[1-(2,3-dioleoyloxy) propyl]-*N,N,N*, trimethylammonium methyl sulphate, a conventional transfection reagent used for DNA transfection. These observations are in agreement with earlier findings from polymer science, stating that the cytotocicity of polymers is proportional to their molecular size.[16]

Modified amino-terminated PAMAM dendrimers based on a trimesyl core show a similar generation-dependent cytotoxicity profile (HeLa cells), albeit these modified PAMAM dendrimers showed significantly lower toxicity in comparison to traditional polycationic carriers as PEI and PLL.[36]

In contradiction to this general trend, dendrimers based on an alkylsilan scaffold grafted with non-toxic hydroxyterminated polyethyleneoxide (PEO) have shown an inverse generation-dependent behaviour, where the lower generation dendrimers are more toxic than the higher generation dendrimers.[25] This supports the notion that, in lower generation dendrimers, a potentially toxic dendrimer core will be more easily exposed for interactions with the surroundings *e.g.* cell surfaces, whereas the core in higher generation dendrimers is more shielded off by a non-toxic surface, resulting in low cytotoxicity.

In addition to the size of the dendrimer, the flexibility of the polymer skeleton also seems to be important for the cytotoxicity of a given polymer. It has been found that amino PAMAM dendrimers, with their lower flexibility and globular structure, are less cytotoxic in comparison to amino-functionalised linear polymers. This may be explained by the lower ability for the less flexible and globular dendrimer to perform sufficient adhesion to cell surfaces.[16]

The flexibility of the PAMAM dendrimer skeleton in amino-terminated dendrimers can be increased by partial fragmentation of the dendrimer by heating or solvolysis.[36,37] The increased flexibility of these fragmented amino-terminated PAMAM dendrimers makes these more prone to complexation with DNA and thereby increase their usefulness as non-viral transfectants. *In vitro* mesasurements carried out on these fragmented dendrimers show that their cytotoxicity is lower compared to that of the perfect dendrimers. Although this seems to contradict the notion that a flexible polymer is considered more cytotoxic than a more rigid polymer, another factor comes into play as well, namely, the density of the surface primary amines. It has been found that, although a certain number of primary amines is required to sufficiently complex the DNA, the amount of primary amines present on the surface of a fragmented dendrimer is lower compared to the intact dendrimers. As a consequence, the overall charge density is decreased, and the less cytotoxic, alkyl shielded, tertiary amines to a larger extent serve as a source for ion pairing with DNA (Figure 2.9).[38]

Also, the haematotoxicity (rat blood cells) of amino-terminated PAMAM dendrimers was shown to be dependent on the dendrimer generation, with increasing haemolytic activity of the dendrimer upon increasing generation.[25] A similar behaviour has been observed for the interaction between human blood cells (erythrocytes, Central Blood Bank, Lodz, Poland) and PAMAM dendrimers.[39] Amino-terminated PAMAM dendrimers generally showed lower haematotoxicity compared to amino-terminated dendrimers of the PPI and PEI type. This may be due to the significantly less-charged PAMAM scaffold consisting of a mix of amides and tertiary amines in comparison to the highly cationic scaffolds found in PPI and PEI dendrimers, being solely based on a tertiary amine skeleton.[25] The PAMAM dendrimer scaffold also has a lower degree of flexibility compared to PPI and PEI, because of the presence of "sp^2 amide carbonyl" carbons with lower flexibility compared to alkyl bonds.

Interestingly, it has been found that polybenzylether-based dendrimers (Fréchet type) covered with "cytotoxicity quenching" carboxylate surface groups show

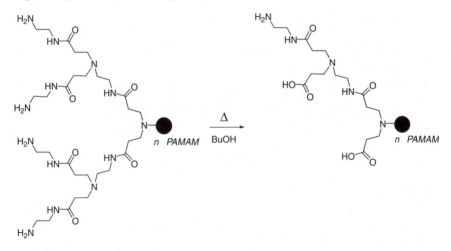

Figure 2.9 *Heat-mediated fragmentation of PAMAM dendrimers increases the flexibility of the dendrimer scaffold. Despite this, the decreased amount of cytotoxic primary amines and the increased amount of non-toxic carboxylic groups at the dendrimer surface leads to a decreased cytotoxicity of the fragmented dendrimer*

haemolytic activity after 24 h incubation with rat blood cells.[25] The membrane-disrupting property of this dendrimer type may be caused by hydrophobic interactions from the dendrimer scaffold with lipid chains in the cell membrane.

As a general rule of thumb, dendrimer scaffolds containing numerous positive charges and/or hydrophobic motifs are cytotoxic; however, their ability to be "neutralised" by *e.g.* non-toxic surface groups may vary depending on the nature of the scaffold.

When comparing two types of dendrimers based on a poly(ether imine) skeleton, the dendrimer with an aromatic core was significantly more cytotoxic (~30% decrease in cell viability, mammalian CHO cells and breast cancer T47D cells) compared to a similar dendrimer based on a diaminoethylene ether core, albeit neither of these dendrimers exhibited pronounced cytotoxicity.[40] Further modification on these poly(ether imine) motifs has been carried out and the cytotoxicity profiles of these dendrimers are promising with respect to biological applications (Figure 2.10).[41]

Recently, dendrimers based on a polyester skeleton and covered with PEG as surface groups have shown to be non-toxic *in vitro*, posing an ideal scaffold for drug delivery or therapeutical applications.[26] At high concentrations (40 mg mL⁻¹) some inhibition of cell growth was seen *in vitro* (murine melanoma cell-line B16F10, sulforhodamine B assay), however no cell death was monitored.

2.4 *In Vivo* Cytotoxicity of Dendrimers

In comparison with the large number of *in vitro* investigations on dendrimers, investigations concerning the *in vivo* toxicity of dendrimers are sparse. The full organism seems to react in a somewhat different way to what is observed by *in vitro* investigations, *e.g.* dendrimers that show cytotoxicity *in vitro* do not necessarily lead to

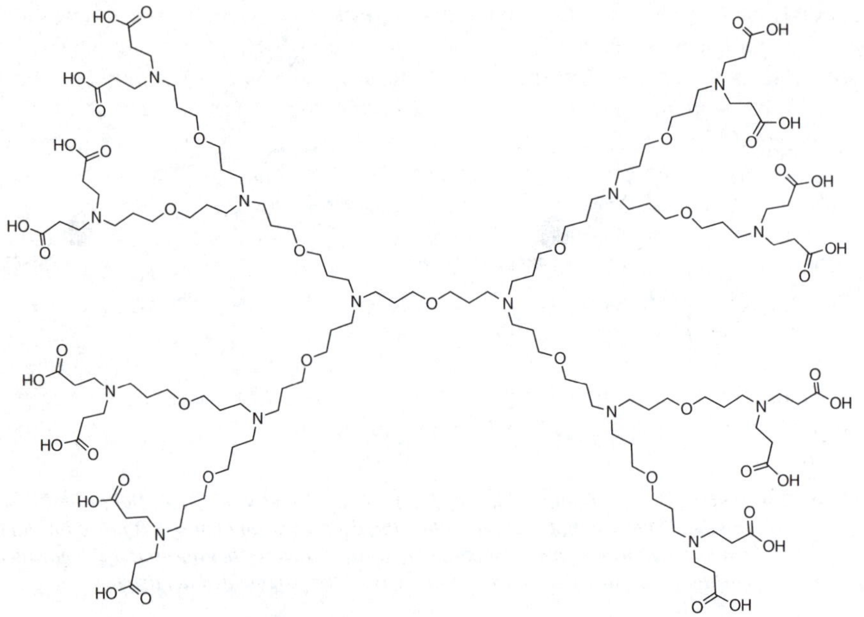

Figure 2.10 *Dendrimer with a low-toxic scaffold published by Krishna and co-workers*[39]

toxicity *in vivo* and *vice versa*. Furthermore, the testing of a certain substance in different animal models may have very different outcome, *e.g.* being highly toxic to one animal and almost non-toxic to another.

Typical *in vivo* symptoms are inflammation around the injection area, granuloma formation or tissue necrosis; other severe symptoms may be hyperthermia, vomiting and weight loss. The more radical LD-50 measures the dose/kg required that is lethal for 50% of the animals tested. Most *in vivo* studies reported on dendrimers so far have been carried out in mice or rats.

Amino-terminated or modified PAMAM dendrimers up to G5 do not appear to be toxic *in vivo*;[34,42] however, administering amino-terminated G7-PAMAM dendrimers to mice (45 mg kg^{-1}) proved fatal (20% of the animals) after 24 h in several experiments.[34] This is also in agreement with later findings where the administration of G7-PAMAM dendrimers to mice leads to the death of 20% of the animals, concluding that the application of G7-dendrimers *in vivo* "seem to have the potential to induce problems".[25] Therefore, high-generation (>G6) PAMAM dendrimers are generally not considered appropriate molecular tools in bioapplications because of their high toxicity.[34]

Melamine-based dendrimers administered i.p. in high doses (160 mg kg^{-1}) resulted in 100% mortality 6–12 h after injection. Lower doses (40 mg kg^{-1}) of melamine-based dendrimers (i.p. administration) appeared to be hepatotoxic to mice after 48 h based on the measurement of liver enzyme activity, and liver necrosis was observed.[14] However, administering low doses of these dendrimers (*e.g.* 2.5 mg kg^{-1}) did not pose any problems *in vivo*.

Polylysine-based dendrimers having sulfonate or carboxylate surface groups show good antiviral activity against Herpes simplex virus (HSV), and have in this regard been tested for toxicity *in vivo* (mice). Preliminary studies of these dendrimers reveal no toxicity *in vivo* in the concentrations used (10 mg mL^{-1}).[42]

Upon applying dendrimer scaffolds built up of non-charged hydrophilic functionalities generally improves the biocompatibility. Dendrimers based on a polyester scaffold with hydroxy or methoxy surface groups have shown to be non-toxic *in vivo* (mice) even at very high concentrations (1.3 g kg^{-1}).[26,43] At very high concentrations (40 mg mL^{-1}) some inhibition of cell growth *in vitro* was observed, but no acute or long-term effects were measured *in vivo* at doses of 0.1 g kg^{-1}. The LD$_{50}$ value of a polyester dendrimer shown in Figure 2.11 was 1.3 g kg^{-1}; the course of death is unclear but was not due to a general haemolytic effect.[26] The polyester scaffold is very interesting from a biological point of view; first, it is non-toxic, second, the ester moiety may be degraded *in vivo* by hydrolytic enzymes after release of a drug, or other bioactive substance.

In vivo studies of triazine-based dendrimers carrying PEG2000 surface groups in mice conclude that this dendrimer type did not have any liver or kidney toxic effect on these animals, measured from the urea nitrogen content (large secretion indicates renal dysfunction) or alanine transaminase liver enzyme levels (alanine transaminase is a liver enzyme associated with activity and viability of the liver cells).[18]

In conclusion, dendrimers generally do not appear to induce significant toxicity *in vivo*, however, some general trends can be put out. The *in vivo* toxicity is dependent on molecular size (generation) of the dendrimer, where large size dendrimers (>G6) seem to be toxic to the organism, especially when the dendrimer in addition has cationic surface groups. In these cases, there is some correlation with the cytotoxicity observed *in vitro*. The *in vivo* observed symptoms may be a consequence of high

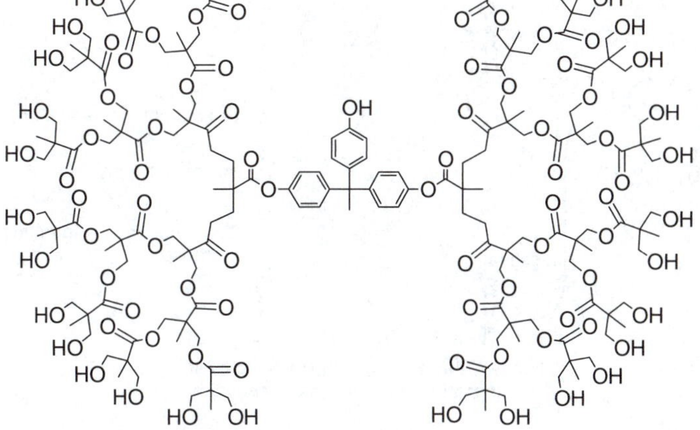

Figure 2.11 *Polyester-based dendrimer for drug delivery purposes. This dendrimer type shows very low in vitro and vivo toxicity*[26]

cytotoxicity in combination with poor biopermeability of these molecules, which retards the normal pathways of excretion from the organism. The following accumulation in the organs leads to tissue necrosis and severe malfunction due to cytotoxicity. Modification of the surface with non-invasive PEG chains on the dendrimer surface improves the properties of the dendrimer *in vivo* towards longer blood half-lives and less accumulation in the organs (*vide infra*).

2.5   Biopermeability of Dendrimers

In order to use dendrimers for drugs or drug delivery, their biopermeability on an intracellular level as well as their ability to cross intestinal barriers (epithelia) or blood-tissue barriers (endothelia) *etc.* has to be taken into consideration. The ability to penetrate various biobarriers depends on the specific function of the biobarrier. The epithelial layer has to be tight in order to filter the large amount of microorganisms *e.g.* bacteria or viruses taken in, in every breath of air, or to be able to carefully filter beneficial compounds from non-beneficial compounds in every meal we eat. In *e.g.* the blood circulation, which transports beneficial compounds such as carbohydrates, proteins, peptides and oxygen, the blood-tissue barrier (endothelial tissue) should be more accessible to penetration in order to ensure facile transport of these vital entities into the tissue cells (Figure 2.12).

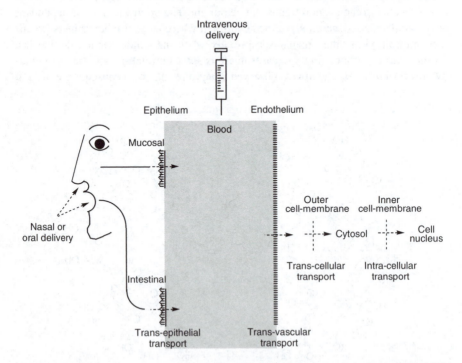

Figure 2.12 *The biobarriers bypassed by a low-molecular compound e.g. a G2-PAMAM dendrimer during uptake in the mammalian organism applied by different delivery pathways*

Therefore, the endothelial cellular network is a more "loosely" bound and porous cellular network in comparison to the epithelial layer. Agents circulating in the blood may extravasate out into the tissue by passive diffusion through these pores, which are permeable to nanosize molecules *e.g.* small proteins, lipids, saccharides *etc.* As cell surfaces are negatively charged and both epithelia and endothelia is built up by cellular networks, the epithelial as well as the endothelial tissues have negatively charged surfaces, and interaction with *e.g.* cationic or anionic dendrimers may to a large extent be governed by electrostatic forces.

For most applications like transfection of drugs or gene delivery into cells, the transfection efficacy, *i.e.* transmembrane permeability, of the gene- or drug carrier should be high. However, for dendrimeric drugs or drug carriers, which have to stay in circulation for the time needed to have the desired effect, the clearance of the organism should not be too fast. In order to investigate the potential of dendrimers as drug- or gene carriers and the use of dendrimers as scaffolds for MR-contrast agents, numerous studies have been carried out *in vitro* concerning the permeability of these adducts across various membranes. These investigations are, despite representing a greatly simplified picture, important tools in the understanding of the biological fate of a certain drug (*vide infra*) (Figure 2.13).

Transcellular and intracellular permeability: For drugs or genetic material to be delivered into the cytosol or cell-nucleus, transmembrane transport is an important factor. As macromolecules, in general, do not have ability to be transported by the transport channels present in cell walls used in the in- or outflux of substrates to and from the cell, the transport across the cell membrane should preferably be by the formation of endosomes, to ensure transport of the substrate or drug into the cell without cell-lysis. As mentioned previously, amino-terminated cationic dendrimers of both PAMAM and PPI type complexed with DNA are less cytotoxic compared to uncomplexed dendrimers. The dendrimer–DNA complexes penetrate the cell membrane via endocytosis, posing a less membrane disruptive and cytotoxic path compared to that of "hole formation".[12] Investigations on the ability of PPI dendrimers

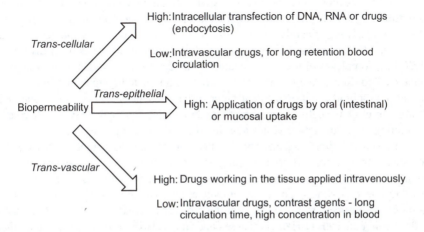

Figure 2.13 *The different demands for permeability of drugs, depending on their function*

to act as good DNA transfectants show that the best transfection efficacy is obtained by the lower generation dendrimers G2-PPI being most potent.[35] This may be explained by the larger flexibility of the lower generation PPI dendrimers compared to the higher generations as well as their higher ability to pass the cellular membrane via endocytosis in comparison to the higher generation dendrimers, which have higher tendency to form holes in the membrane. For PAMAM dendrimers, the permeability of dendrimer–DNA complexes may also be greatly improved by increasing the flexibility of the dendrimer by fragmentation.[37,38] These partly fragmented dendrimers have a larger molecular flexibility to form tight complexes with DNA. In addition, this flexibility gives better possibility to get into close contact with the cell membrane, in contrast to intact dendrimers, which have a more rigid spherical structure.

Moderate amounts of β-cyclodextrin (β-CD) added to G6-PAMAM dendrimer–DNA complexes have also been shown to enhance the transfection efficacy. By the addition of β-CD, the DNA–dendrimer complex composition is altered leading to larger permeability of these complexes.[44] In general, the spherical shape of dendrimers is not ideal for permeation through cell membranes because of the limited ability to get in close contact with the membrane. This is because of limited flexibility and because the formation of supramolecular dendrimer aggregates may lead to membrane-disruptive interactions instead.

Polylysine dendrimers complexed with DNA can obtain enhanced cell membrane permeability by PEGylation of the dendrimer surface.[27] However, the transfection efficacy of the polylysine dendrimers were lower in comparison to DNA complexes with the linear polymers, which may be explained from the lower flexibility and more spherical shape of the dendritic molecules. These results oppose the expectations that polylysine dendrimers would perform better as transfection agents owing to their higher cellular uptake in comparison to *e.g.* PEI polymers or DOTAP, and the authors conclude that the cellular uptake of the transfectant is not the limiting factor in gene delivery.[27]

Modification with lipid chains on the surface of PAMAM dendrimers results in increased permeation and reduced cytotoxicity compared to the unmodified PAMAM dendrimers.[45] For lysine-containing dendrimers, the incorporation of palmitoyl chains similarly significantly enhances the ability to transport DNA and genetic constructs into human epithelial carcinoma (HeLa) cells and mouse skeletal myoblasts (C2C12). Complexes of the palmitoylated lysine dendrimer with DNA showed to be tighter complexed in comparison to the non-palmitoylated lysine dendrimer.[46] The presence of a lipid chain on the dendrimer surface enhances the interaction and penetration through the cellular membranes, and may therefore play a favourable role in promoting the formation of vesicles during the endocytosis process.

The favourable lipid–cell membrane interaction in the endocytotic DNA uptake, together with the lower cytotoxicity compared to cationic dendrimer carriers, has lead to intense investigations on various designs of dendrimers with an incorporated lipid part. Lysine dendrimers having an amphipathic asymmetrical design have shown good *in vitro* properties in transfection of a reporter plasmid (pSVβgal) into hamster kidney epithelial cells (BHK-21). These dendrimers also showed to be less toxic than earlier reported dendrimer carriers for plasmid delivery (Figure 2.14).[47]

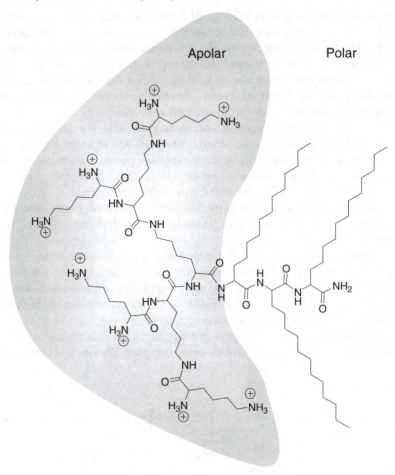

Figure 2.14 *Amphipathic dendrimer published by Shah and co-workers[47] for a more effective gene delivery to cells together with a low cytotoxicity*

Transepithelial permeability: In order to investigate the migration of a given compound through an epithelium some frequently used *in vitro* systems will briefly be mentioned here. The cellular monolayers often consist of Caco-2 (human colon adenocarcinoma) cells or Madin–Darby canine kidney (MDCK) cells, a cellular system originally derived from male Cocker Spaniel distal renal epithelium, with similar properties to Caco-2 cells. Both cellular systems have the beneficial properties of being able to form uniform monolayers bound together by cell–cell tight junctions within a few days.[48] However, these monocellular systems are very simplified models for the epithelial barrier, consisting of different cells, basal membranes and supporting tissues (mucosal secretion tissue *etc.*), which varies depending on the type of epithelial barrier.

To get a more "full" picture of the transport through the epithelial layer in the intestines the "everted intestinal sac system" is sometimes applied. In this system, part of the gut is everted to form a sac, where the flux through the wall (epithelial layers) of the sac is measured. The complexity of this system in comparison to the epithelial cellular monolayer may result in diverging results obtained in comparing these systems. The everted intestinal sac system comprises several different cellular systems with different permeation properties, whereas the monolayer comprises only one cell type. Furthermore, several size and charge-specific transport systems may come into play when applying the everted sac system.

The crossing of a given compound through the epithelium or a cellular monolayer may proceed by two pathways, the *transcellular pathway*, in which the compounds penetrates the cell via endocytosis, or the *paracellular pathway* where the compound is transported via the tight junctions between the cells without penetration of cell membranes (Figure 2.15). The paracellular transport is more sensitive to the molecular dimensions and weight of the polymer, where the larger polymers show a lower paracellular permeability compared to the polymers having lower molecular weight and smaller molecular dimensions.[49]

The molecular migration through the epithelial layer may proceed in two directions, either from the apical surface, facing the lumen, in direction towards the basal (or basolateral) surface (A→B), or migration may occur in the opposite way from the basal to the apical surface (B→A). The A→B permeability should be high for orally delivered drugs to have a maximum delivery to the organism. However, B→A pathways may be favoured, when systems like the ABC or D-glycoprotein (D-gp) efflux transport systems in the intestine come into play. The epithelial layer has, as

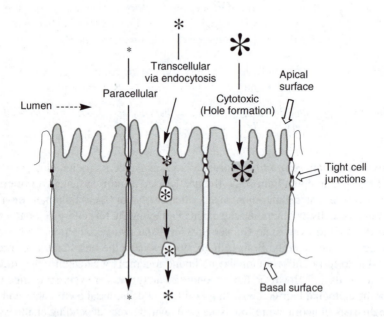

Figure 2.15 *Different pathways for the apical to basal transport of different generations (size) of cationic dendrimers (shown as asterisks) across the intestinal epithial cell layer*

an evolutionary precaution, several efflux systems for the secretion of unwanted compounds back into lumen by the B→A pathway. The G-glycoprotein diminishes absorption and oral bioavailability of potential drugs by serving as an efflux transporter, secreting the drug out of the epithelial layer into the gut lumen.

Initial studies performed on Caco-2 cell monolayer as epithelial model showed that lower generation (G1–G3) cationic PAMAM dendrimers had higher permeability in comparison to higher generation dendrimers, which were also hampered by their increasing cytotoxicity. Generally, the B→A permeability of these cationic PAMAM dendrimers is significantly higher compared to the A→B permeability; however, by adding lipophilic compounds *e.g.* palmitoyl carnitine, the A→B transport through the epithelial cells is significantly increased.

As positively charged polymers have shown enhanced paracellular permeability via loosing up the intercellular tight junctions it has been rationalised that the permeation of these lower generation cationic dendrimers through the epithelium takes place predominantly by paracellular transport. As a consequence of the ability for cationic dendrimers to modulate or loosen the tight intercellular junctions, they can serve as additives to increase the permeability of small molecules like mannitol by the paracellular route.[49]

Later studies on the same system revealed that the transport of amino-terminated PAMAM dendrimers (G3–G5) through the epithelium occurs both by paracellular and transcellular pathways as the transepithelial permeability decreased when adding the endocytosis inhibitor colchicine.[50] Recent studies on G4-PAMAM dendrimers using flow cytometry as well as transmission electron microscopy further support the transepithelial transport of medium generation dendrimers (G4) via a predominantly transcellular route. These studies show that a significant amount of dendrimer is found inside the cells during transport, and in addition, the permeability is decreased by the addition of colchicines.[44]

In the Caco-2 cell monolayer systems, studies on anionic carboxy-terminated PAMAM dendrimers (G3.5 and G4.5) show lower A→B permeability compared to amino-terminated dendrimers.[50] The lower permeability of the anionic dendrimers is in good agreement with earlier findings, showing that anionic compounds like folate or lactate have significantly lower transepithelial permeability in comparison to cationic compounds like methylamine, being able to interact favourably with the negatively charged epithelial surfaces.[51] However, investigations show that mid-generation (G3.5, G4.5.) anionic carboxy PAMAM dendrimers have an enhancing effect on transepithelial transport of *e.g.* mannitol by the paracellular pathway. This indicates that these dendrimers may also have the ability to loosen the intercellular junctions, however, without being able to cross these junctions.[23]

The A→B permeability as well as the toxicity profile of the amino-terminated PAMAM dendrimers could be improved towards significantly higher permeability and lower cytotoxicity by partial lipidation (lauroylation) of the amino surface groups with lauroylchloride. As with the unmodified cationic dendrimers, inhibition or facilitation of penetration was observed by addition of endocytosis inhibitor or EDTA, respectively. This indicates that transepithelial transport of the lauroyl dendrimers takes place by transcellular transport, where the lipid chains have a beneficial effect on the endocytotic cellular uptake.[44]

In contrast, it has been found that non-charged small PEG polymers cross the rabbit colonic epithelium exclusively by the paracellular pathway, which is in agreement with the expected non-invasive behaviour of PEG chains towards the cellular membranes.

For drugs having poor A→B permeability or which may serve as substrates for the P-glycoprotein (P-gp) efflux system, the A→B transepithelial transport has been significantly enhanced by covalent conjugation of these drugs to dendrimers.[52] Conjugation of the P-gp substrate propranolol (β-blocker used in the treatment of hypertension) to either unmodified cationic or lauroyl PAMAM dendrimers (G4) increased the A→B permeability in comparison to non-conjugated propranolol. In addition to the better permeability, the solubility of the sparingly soluble propranolol was increased by conjugation to the dendrimer. Non-covalent mixtures of dendrimers with propranolol did not increase transepithelial permeability of the drug.

A facile renal excretion of dendrimer-based agents out of the body is crucial in order to apply these macromolecules as scaffolds having minimal toxic effects on the organism. *In vitro* studies on the transepithelial transport of amino-terminated PAMAM dendrimers (G1–G5) in MDCK cells showed that the G5-PAMAM dendrimer possessed the largest permeability; however, in these studies no clear linear dependence between the dendrimer generation and permeability was found.[48] The A→B permeability of cationic amino-terminated PAMAM dendrimers across MDCK cells was shown to follow the order.

G5-PAMAM >> G2-PAMAM ~ G1-PAMAM > G4-PAMAM > G3-PAMAM

When applying the everted intestinal sac system in *in vitro* investigation of transepithelial transport of dendrimers, somewhat different results are obtained compared to the "ideal" cellular monolayers. Investigations *in vitro* using an everted rat intestinal sac show that the anionic carboxylate-terminated PAMAM dendrimers rapidly cross into the intestine with a faster transfer rate than cationic dendrimers. The high transfer efficacy of these dendrimers could make these good potential candidates as building-blocks in oral delivery systems.[53]

Extravascular (transendothelial) permeability: In the internal vascular barriers (endothelia) separating the blood circulation from the surrounding tissue, modulation of the molecular properties of drugs or drug vehicles is important to ensure a proper transendothelial transport (extravascular permeability) into the tissue, or in the case of intravascular drugs a prolonged retention of the drug in the vascular system. The previously mentioned epithelial layer is the "boundary" against the surroundings and therefore should be able to exhibit a discriminative uptake of compounds before further metabolic processing. In contrast, the vascular endothelia should be permeable to a large variety of useful agents (*e.g.* carbohydrates, proteins), which need to be transported from the blood stream into the surrounding tissues. As a consequence, the microvascular endothelium cells are not as tightly joined together and contains numerous small endothelial pores with radii of ~5 nm, which makes this "biobarrier" permeable to smaller molecules.[54] Dendrimers and hydrophilic solutes in general diffuses through the endothelial wall by "restricted diffusion", where the larger molecules, due to their slower translational motion and larger extent of exclusion from the micropores, experiences longer extravasation times. The three-dimensional conformation of the polymer strongly affects the diffusion process through the endothelial wall. Comparative studies between globular and linear polymers of similar molecular weight and hydrodynamic volumes generally conclude that the globular polymers will have slower transendothelial diffusion compared to the linear molecule. Hence the globular polymers are more dependent on the property of the endothelial pores, such as pore diameter and homogenicity of the pores. This limits the transendothelial efficacy of globular polymers (*e.g.* dendrimers) for the use as contrast agents for the bypassing of diseased endothelia *e.g.* in tumors, which have a more heterogeneous porosity in comparison to normal endothelial barriers.[55]

For amino-terminated cationic PAMAM dendrimers (G1–G5), *in vivo* studies indicate that the time of transport from the blood vessels into the tissue (extravasation time) increases with increasing generation and molecular weight of the dendrimer.[54]

Extravascular permeability:

G1-PAMAM > G2-PAMAM > G3-PAMAM > G4-PAMAM > G5-PAMAM

In the design of MRI (magnetic resonance imaging) intravascular contrast agents, a delicate balance between extravasation and the ability to be secreted out of the body comes into play. The extravasation should not be too high in order to give a detailed visualisation of the vascular system. Consequently, these contrast agents should have relatively large molecular dimensions to retard the extravasasive (transendothelial) transport into the surrounding tissues. On the other hand, contrast agents based on too large molecular scaffolds will be retarded in the renal excretion from body. If the contrast agent is retained for too long time before clearance from the body, the leakage of toxic metal ions from the agent may increase, which is not desirable from a clinical point of view. However, sometimes fast body clearance of a contrast agent is desirable *e.g.* in perfusion measurements based on repeated investigations.

Subtle structural differences in the scaffold of dendrimers of similar molecular dimensions may affect the blood clearance and excretion profiles quite significantly. Comparative studies performed between similar-sized dendrimer gadolinium contrast agents based on G7-PAMAM (starburst) (192 surface amines) and G7-PAMAM with an ethylene diamine (256 surface amines) core show that the glomerular filtration (filtration through the kidneys) proceeded faster with G7-PAMAM (starburst). This starburst contrast agent experienced a relatively fast clearance (extravasation) from the blood and renal accumulation in comparison to the PAMAM agent. As the PAMAM agent is able to bind more Gd (III) per molecule and was retained longer in the blood stream, this agent visualised the fine blood vessels for a longer time, which is important for intravascular investigations. On the other hand, as a consequence of higher renal accumulation, the starburst-based agent gave brighter images of the kidneys in comparison with the PAMAM agent.[56]

Also, in comparing PAMAM and PPI dendrimers having identical numbers of surface amino groups (and generation), the lower molecular weight PPI dendrimers have a faster body clearance in addition to a slow extravasation, which makes this class of dendrimers interesting as scaffolds for contrast agents used in the clinic.

Surface modification of dendrimer-based contrast agents alters the extravasation profile of the contrast agent. Partial PEGylation of the surface amines of PAMAM dendrimer contrast agents prolongs intravascular retention and gives less accumulation in the organs. Opposite to the positively charged terminal amines, the uncharged PEG chains have a low affinity towards receptors and negative charges on the vascular walls, resulting in a decreased binding to the vascular endothelium and following extravasation.[57] In addition to longer blood circulation times, the low affinity and non-invasive behaviour of PEG also leads to a more rapid excretion through the kidneys and low accumulation in the organs.

Molecular motifs with low extravascular permeabilities are important tools in the development of intravascular drugs (thrombin inhibitors *etc.*) that should be retained in the blood circulation for a prolonged period. As the extracellular vascular tissue comprises a large content of negatively charged glycosaminoglycans (GAGs), it has the ability to bind cationic molecules as *e.g.* amino-terminated dendrimers, and these cationic molecules may therefore be important delivery tools for localised delivery of intravascular drugs at the vascular wall.

As the transendothelial extravasation time increases with increasing molecular size, dendrimers serving as intravascular pharmaceutical agents should have relatively large molecular dimensions in order to be retained in the vascular system. Studies *ex vivo* on a rat carotid artery segment show that G5- and G6-PAMAM (starburst) dendrimers partially modified with FITC had much longer intravascular retention times in comparison to GAG binding peptides.[58] The PAMAM dendrimers could be accumulated in the vascular tissue and retained at the vascular wall for several days *in vivo*.[59] This ability makes the unmodified higher generation cationic dendrimers promising delivery agents for the retention of drugs in the vascular system.

In conclusion, some general trends can be setup for the permeability of dendrimers through the various membranes. In transmembrane transport of dendrimer/DNA complexes into the cell, there is a correlation between size and the ability to cross the membrane via endocytosis. Complexes containing low-generation dendrimers (<G4) penetrate the cell membrane predominantly through endosomal formation (endocytosis) without compromising the structural integrity of the cell, whereas DNA complexed with higher generation dendrimers (>G4) lead to a higher degree of hole formation and lysis of the cell.

Transepithelial transport of dendrimers in monolayers of epithelial cells (*in vitro*) shows that cationic dendrimers have larger ability to be transported through the monolayer than anionic dendrimers (G2.5-PAMAM and G3.5-PAMAM), which show a very low apical to basal permeability. The poor permeability of the anionic dendrimers in monocellular systems may be attributed to their disability to get in close contact with the negatively charged cell walls and intercellular junctions (charge repulsion), which is crucial in the transport through the epithelial cell layer. The transport of cationic dendrimers through the cellular monolayers is generation dependent, the lower generations showing higher permeability compared to higher generation dendrimers. However, the full picture of generation dependence is blurred because of the increasing cytotoxicity of the high-generation dendrimers. Small cationic dendrimers (G1–G3) seem to be transported by both paracellular and transcellular routes, with increasing paracellular transport upon decreasing generation (size). In contrast, dendrimers of slightly higher generation (G2–G4) is transported more frequently by transcellular transport. For higher generation cationic dendrimers, the transcellular (endocytotic) pathway is hampered by membrane hole formation and subsequent cell-lysis (cytotoxicity). Both with respect to transmembrane, transcellular and paracellular penetration, partial modification of the dendrimer surface by lipidation greatly improves the permeability and reduces cytotoxicity of the cationic PAMAM dendrimers proportional to the number of lipid chains introduced.[50] The opposite order of permeability of cationic *versus* anionic dendrimers shows in more complex *in vitro* systems like the everted intestinal sac. Here, the anionic dendrimers have higher ability to be transported through the epithelial wall, leading to higher intestinal uptake. This implies that transepithelial transport of dendrimers is not performed only by paracellular and transcellular transport, which require closer contact with anionic cell walls, but that other factors *e.g.* anionic specific transport systems may play a role as well. Also, in this system the lower generation anionic dendrimers generally have higher permeabilities compared to higher generation dendrimers (G6.5), which to a larger extent accumulated in the tissue, indicating that this transport is dependent on size and conformation of the dendrimer.

The transport of cationic dendrimers and other molecules through the endothelial wall happens through size-dependent restrictive diffusion, where the lower generation dendrimers show higher transendothelial permeability *in vivo* compared to dendrimers of higher generation. High-generation cationic dendrimers may due to their poorer transendothelial permeability be utilised as "stickers" to the vascular walls and endothelia by ionic interaction between the cationic dendrimer and the anionic vascular wall. These dendrimeric stickers may be used for retaining drugs in the vascular system. Modification of the dendrimer surface by PEGylation to give a more non-charged surface decreases the transvascular permeability, achieving longer circulation in the blood system.

2.6   Biodistribution of Dendrimers

The molecular structure and the molecular dimensions of a dendrimer are also important factors in how dendrimers distribute in the body. As a rule of thumb, the medically applied macromolecule should have a molecular weight larger than 20 kDa to act as a blood pool agent, that stays in circulation for a prolonged period. On the other hand, the molecule should have a mass smaller than 40 kDa to be able to be secreted through the kidneys, which is the preferred route of secretion.[60] Furthermore, the surface of the macromolecule should be compatible with the various bioadministration approaches. Polycationic species generally could pose a health risk for the organism because of their cytotoxicity, furthermore caution should be taken in the delivery of these compounds to the organism, *e.g.* polycationic compounds administered through the lungs are known to cause severe lung oedema and high mortality in rats.[61] Also, the nature of the surface group greatly affects the distribution of a dendrimer in the body. Initial studies on PAMAM (starburst) dendrimers having cationic surface groups found that this type of dendrimer was predominantly excreted through the kidneys to the urine, the high-generation PAMAM dendrimer (G7) being cleared faster from the body via the kidneys.[34] This is in good accordance with the fact that the molecular conformation of high-generation dendrimers is significantly different from low-generation dendrimers, and that these large, less-flexible, spherical structures have a reduced ability to interact with biological membranes by *e.g.* extravasation into tissues, paracellular transport or endocytotic uptake into cells. Later investigations have revealed that cationic amino-terminated PAMAM and PPI dendrimers are quickly cleared from the blood circulation, albeit high levels of dendrimer were accumulated in the liver and kidneys after a 4-h period from administration. Also, anionic PAMAM dendrimers accumulate in the liver. However, these dendrimers have longer circulation times compared to their cationic counterparts, possibly due to the previously mentioned poorer transendothelial permeability. Modification of the dendrimer surfaces with PEG is generally performed to modulate the biodistribution properties towards longer circulation times (blood half-life), and lower amount of accumulation in the liver. Long blood circulation times are desirable properties for *e.g.* contrast agents, and may also diminish the toxic effects seen upon accumulation in the organs.[60] Several reports indicate that modification of the dendrimer surfaces with PEG creates dendrimeric agents with diminished side effects such as cytotoxicity and inflammatory reactions in the body.[50,56,57]

Polylysine dendrimers having lipid surface groups have been applied to Sprague–Dawley rats and shown to be predominantly accumulated in the small and large intestines before migrating to the different organs or cleared from the body. These types of dendrimers predominantly accumulate in the liver and in blood.[62]

As modification of the cationic surface groups on a dendrimer generally leads to less toxicity, the modification of amino-terminated dendrimers with surface groups having reduced polarity has been given some attention. Methylation of amino-terminated PAMAM dendrimers creates an apolar dendrimer surface, which leads to altered distribution properties in comparison to the unmodified dendrimers, with high accumulation in the pancreas. It has been suggested that such methylated dendrimers could therefore serve as pancreas-specific drug delivery devices.[34]

Other investigations on dendrimers with uncharged surfaces have been carried out in mice with ^{3}H-labelled G6-PAMAM dendrimers either with unmodified cationic surfaces or acetylated non-charged surfaces. These studies show that the change in surface charge alters the biodistribution of the respective dendrimers. The cationic PAMAM dendrimer to a larger extent deposited in the tissues compared to the PAMAM dendrimer with a non-charged surface. Fast body clearance was observed by administration of both dendrimer types. The distribution of the different dendrimers in the organs was similar to the highest levels found in the lungs, liver and kidneys, and the lowest levels were found in the brain. Medium levels were found in the heart, pancreas and the spleen.[63]

In conclusion, the faster transport of the cationic dendrimers from the blood system into the tissues may be rationalised from the higher affinity of the positively charged molecule towards the vascular wall (*vide supra*). The body clearance time is dependent on the molecular size (generation) of the dendrimer. High-generation cationic dendrimers show fast body clearance, and organ accumulation, whereas anionic dendrimers differ from the cationic dendrimers in having longer circulation times. Long circulation times (blood half-lives) are observed with dendrimers having anionic or non-charged (*e.g.* PEGylated) surfaces, because of the lower ability to bind to the vascular walls for subsequent transport into the tissues.

2.7 Immunogenicity of Dendrimers

When looking at unmodified dendrimers having amino surface groups, there are some contradictionary results regarding their immunogenicity. Initial reports on the immunogenicity of dendrimers showed low or modest immunogenicity of unmodified amino-terminated PAMAM dendrimers (G3–G7).[33,34] However, Kobayashi and co-workers[56] later found some immunogenicity of unmodified PAMAM dendrimers of similar generation number. Investigation of high-generation cationic PAMAM dendrimers as well as high molecular weight PEI and polylysine polymers has shown the ability to activate the complement system (see also Chapter 4). However, complexation with DNA greatly reduced the complement-activating ability of these polymers, possibly due to charge reduction.[64] PEGylation of the amine surface groups significantly reduces the immunogenicity of the dendrimer.[56] The introduction of PEG chains generally results in highly hydrated surfaces preventing *e.g.* protein denaturation and clothing, which may lead to activation of the immune system. Interestingly, antiimmunogenic effects may be obtained by partly covering an anionic PAMAM dendrimer with the carbohydrates glucosamine or glucosamine 6-sulfate (Chapter 4). Such derivatives quench immunogenicity and have an antiinflammatory effect on scar tissue formation. This antiinflammatory effect is accredited to inhibition of Toll-like receptor 4 (TLR 4) present on monocytes and following down-regulation of inflammatory cytokines (*e.g.* TNFα) which may otherwise lead to excessive scar tissue formation.[65]

Studies performed on mice show that amino-terminated PAMAM dendrimers may work as adjuvants (immunostimulating compounds) when injected together with ovalbumin.[32]

In conclusion, dendrimers have varying degrees of immunostimulatory properties depending on their surface functionalities. As a general trend, the immunogenicity of unmodified amino dendrimers is low, but seems to increase with increasing generation of the dendrimer.[57,64] The increasing cytotoxicity of the high-generation dendrimers may also lead to inflammatory responses from the organism, and high-generation dendrimers have found to activate the complement system in the host. Several surface functionalities have been found to have a suppressive effect on the immunogenicity of dendrimers, *e.g.* PEG, hydroxyls and certain carbohydrates. As we shall see in Chapter 4, highly immunogenic dendrimers *e.g.* for vaccines or adjuvants may be created by introduction of *e.g.* T-cell epitopes or antigenic peptides as surface groups.

2.8 Summary

The multivalent nature of the dendritic motif is ideal in the amplification of binding in biological systems dominated by weak binding affinities, *e.g.* carbohydrate–protein interactions. In these systems dendrimers, with their ability to expose a multivalent surface together with a simple molecular structure, are promising building blocks in *e.g.* development of vaccines and therapeutics. Several studies on the polyvalent interactions between dendrimers and various polyvalent receptors show an amplification of binding affinity that is not only additive, but also synergistic with a strong increase in binding per binding site of the dendrimeric ligand.

The biological properties of dendrimers are to a large extent determined by the nature of the surface groups, opening up for the possibility to specifically tailor the dendrimers for the desired biological effects. Although being less important, the toxicity of the dendrimer scaffold may also play a role on the toxicity of a dendrimer, where *e.g.* an apolar scaffold may induce cytotoxicity, although carrying surface groups that are non-toxic.

Dendrimers with unshielded cationic surface groups (*e.g.* primary amines) generally show a generation-dependent high cytotoxicity *in vitro*, where there is proportionality between cytotoxicity and generation number. Cationic dendrimers are better tolerated *in vivo*, where only the high-generation (>G6) amino-terminated dendrimers seem to pose problems. The biopermeabilities of unmodified amino-terminated dendrimers are, in contrast to *in vitro* observations, not very good *in vivo*. *In vivo* administration of high-generation (G7) cationic dendrimers generally results in accumulation in the organs, and in some cases severe organ damage presumably due to the high cytotoxicity of these compounds. In addition, the inflammatory effects of high-generation cationic dendrimers *in vivo* lead to immunogenic reactions from the complement system of the host. Whereas the cationic dendrimers alone may give problems *in vivo* and *in vitro*, the charge reduction and/or shielding of the numerous positive charges by complexation with various substances significantly reduces the toxicity of this class of dendrimers. This opens up to the use of cationic dendrimers as carriers for anionic substrates, *e.g.* DNA. The anionic nature of DNA disables close contact and penetration of the negatively charged cell membranes, which is important in therapeutic DNA transfer or gene delivery into the cell. Several studies show that lower generation cationic dendrimers (<G5) both of PPI and PAMAM type show good properties as DNA transfectants *in vitro* and *in vivo*, and that the transfection efficacy of PAMAM dendrimers can be improved significantly by partial fragmentation of the dendrimer.

In contrast to the cationic dendrimers, anionic dendrimers, in most cases, show no toxicity *in vivo* and *in vitro*. This may be rationalised from the fact that charge repulsion prevents close contact between the negatively charged cellular membranes and the dendrimers, avoiding disruptive interaction with the cell membrane and cell-lysis. *In vivo* and *in vitro* results on biopermeability through epithelial and endothelial tissues show good permeability of anionic PAMAM dendrimers, where the lower generation dendrimers show the highest permeabilities. There are contradictionary results when going to very simplified monocellular systems (monolayers), which show a poor biopermeability of anionic dendrimers. However, in these systems, which lack natural transport systems, the charge repulsion between the dendrimer and the cell membrane becomes an important factor. In epithelial monolayers, the anionic dendrimers although not being able to penetrate the wall themselves, show the ability to increase the permeability of low-molecular compounds, presumably by loosing up the intercellular junctions. Furthermore, the more non-invasive behaviour of the anionic dendrimers towards the vascular endothelia results in longer blood circulation times *in vivo*.

Upon converting cationic dendrimer surfaces to a non-charged surfaces by *e.g.* lipidation or PEGylation, significant reduction in cytotoxicity is observed. Contrast agents based on PEGylated dendrimers show long blood circulation times together with high-excretion rates through the organs, as a result of low surface charge together with the ability for the PEGylated surface to be hydrated by the surroundings. In contrast to PEGylated dendrimers, the lipidated dendrimers show good ability to cross the cell membrane and may constitute a promising class of transfection agents. The transepithelial transport of lipidated dendrimers takes place predominantly via the transcellular route, where small PEG-based polymers are transported through the epithelium by paracellular transport. PEGylation of the dendrimers surface reduces the immunogenicity of the dendrimer agent significantly, whereas PLL dendrimers covered with lipid groups show good immunogenic and adjuvant properties.

References

1. M. Mammen, S.-K. Choi and G.M. Whitesides, *Angew. Chem. Int. Ed.*, 1998, **37**, 2754.
2. P.W. Atkins, *Physical Chemistry*, 4th edn, Oxford University Press, Oxford, 1990.
3. P.H. Ehrlich, *J. Theor. Biol.*, 1979, **81**, 123.
4. J.B. Corbell, J.J. Lundquist and E.J. Toone, *Tetrahedron Asymmetry*, 2000, **11**, 95.
5. D. Pagé, D. Zanini and R. Roy, *Bioorg. Med. Chem.*, 1996, **4**, 1949.
6. D. Pagé and R. Roy, *Bioconjugate Chem.*, 1997, **8**, 714.
7. T.K. Dam, R. Roy, S.K. Das, S. Oscarson and C.F. Brewer, *J. Biol. Chem.*, 2000, **275**, 14223.
8. J.J. Lundquist, S.D. Debenham and E.J. Toone, *J. Org. Chem.*, 2000, **65**, 8245.
9. N. Nagahori, R.T. Lee, S.-I. Nishimura, D. Pagé, R. Roy and Y.C. Lee, *Chembiochem*, 2002, **3**, 836.
10. F. Wroblewski and J.S. La Due, *Proc. Soc. Exp. Biol. Med.*, 1955, **90**, 210.
11. T. Mosmann, *J. Immunol. Meth.*, 1983, **65**, 55.
12. S. Hong, A.U. Bielinska, A. Mecke, B. Keszler, J.L. Beals, X. Shi, L. Balogh, B.G. Orr, J.R. Baker and M.M.B. Holl, *Bioconjugate Chem.*, 2004, **15**, 774.
13. N. Karoonuthaisiri, K. Titiyevskiy and J.L. Thomas, *Colloids and Surfaces B, Biointerfaces*, 2003, **27**, 365.
14. M.F. Neerman, W. Zhang, A.R. Parrish and E.E. Simanek, *Int. J. Pharm.*, 2004, **281**, 129.
15. G.A. Brazeau, S. Attia, S. Poxon and J.A. Hughes, *Pharmacol. Res.*, 1998, **15**, 680.
16. D. Fischer, Y. Li, B. Ahlemeyer, J. Krieglstein and T. Kissel, *Biomaterials*, 2003, **24**, 1121.
17. P. Ferutti, S. Knobloch, E. Ranucci, E. Gianasi and R. Duncan, *Proc. Int. Symp. Controlled Rel. Bioact. Mater.*, 1997, 45.
18. H.-T. Chen, M.F. Neerman, A.R. Parrish and E.E. Simanek, *J. Am. Chem. Soc.*, 2004, **126**, 10044.
19. H.-H. Chung, G. Harms, C.M. Seong, B.H. Choi, C. Min, J.P. Taulane and M. Goodman, *Biopolymers*, 2004, **76**, 83.
20. R. Jevprasesphant, J. Penny, R. Jalal, D. Atwood, N.B. McKeown and A. D'Emmanuele, *Int. J. Pharm.*, 2003, **252**, 263.
21. P. Ferruti, S. Knobloch, E. Ranucci, R. Duncan and E. Gianasi, *Macromol. Chem. Phys.*, 1998, **199**, 2565.

22. S. Fuchs, T. Kapp, H. Otto, T. Schöneberg, P. Franke, R. Gust and A.D. Schlüter, *Chem. Eur. J.*, 2004, **10**, 1167.

23. M. El-Sayed, M. Ginski, C.A. Rhodes and H. Ghandehari, *J. Bioactive Compat. Polym.*, 2003, **18**, 7.

24. S. Higashiyama, M. Noda, M. Kawase and K. Yagi, *J. Biomed. Mater. Res.*, 2003, **64A**, 475.

25. N. Malik, R. Wiwattanapatapee, R. Klopsch, K. Lorenz, H. Frey, J.W. Weener, E.W. Meijer, W. Paulus and R. Duncan, *J. Controlled release*, 2000, **65**, 133.

26. O.L. Padilla de Jesus, H.R. Ihre, L. Gagne, J.M. Frechet and F.C. Szoka Jr., *Bioconjugate Chem.*, 2002, **13**, 453.

27. M. Männistö, S. Vanderkerken, V. Toncheva, M. Elomaa, M. Ruponen, E. Schacht and A. Urtti, *J. Controlled Release*, 2002, **83**, 169.

28. H. Yoo and R.L. Juliano, *Nucleic Acid Res.*, 2000, **28**, 4225.

29. P. Rajananthanan, G.S. Attard, N.A. Sheikh and W.J. Morrow, *Vaccine*, 1999, **17**, 715.

30. I. Gössi, L. Shu, A.D. Shclüter and J.P. Rabe, *J. Am. Chem. Soc.*, 2002, **124**, 6860.

31. C.L. Gebhart and A.V. Kabanov, *J. Controlled release*, 2001, **73**, 401.

32. J.M. Andrews, G.C. Newbound and M.D. Lairmore, *Nucleic Acid Res.*, 1997, **25**, 1082.

33. J. Haensler and F.C. Szoka Jr., *Bioconjugate Chem.*, 1993, **4**, 372.

34. J.C. Roberts, M.K. Bhalgat and R.T. Zera, *J. Biomed. Mater. Res.*, 1996, **30**, 53.

35. B.H. Zinselmeyer, S.P. Mackay, A.G. Schatzlein and I.F. Uchegbu, *Pharmacol. Res.* 2002, **19**, 960.

36. X.-Q. Zhang, X.-L. Wang, S.-W. Huang, R.-X. Zhuo, Z.-L. Liu, H.-Q. Mao and K.W. Leong, *Biomacromolecules*, 2005, **6**, 341.

37. M.X. Tang, C.T. Redemann and F.C. Szoka Jr., *Bioconjugate Chem.*, 1996, **7**, 703.

38. M.X. Tang and F.C. Szoka, *Gene Ther.*, 1997, **4**, 823.

39. D.M. Domański, B. Klajnert and M. Bryszewska, *Bioelectrochemistry*, 2004, **63**, 189.

40. T.R. Krishna and N. Jayaraman, *J. Org. Chem.*, 2003, **68**, 9694.

41. T.R. Krishna, S. Jain, U.S. Tatu and N. Jayaraman, *Tetrahedron*, 2005, **61**, 4281.

42. N. Bourne, L.R. Stanberry, E.R. Kern, G. Holan, B. Matthews and D.I. Bernstein, *Antimicrob. Agents Chemother.*, 2000, **44**, 2471.

43. H.R. Ihre, O.L. Padilla De Jesus, F.C. Szoka and J.M. Fréchet, *Bioconjugate Chem.*, 2002, **13**, 443.

44. B.J. Roessler, A.U. Bielinska, K. Janczak, I. Lee and J.R. Baker Jr., *Biochem. Biophys. Res. Commun.*, 2001, **283**, 124.

45. R. Jevprasesphant, J. Penny, D. Attwood and A. D'Emmanuele, *J. Controlled Release*, 2004, **97**, 259.

46. G.P. Vlasov, V.I. Korol'kov, G.A. Pankova, I.I. Tarasenko, A.N. Baranov, P.B. Glazkov, A.V. Kiselev, O.V. Ostapenko, E.A. Lesina and V.S. Baranov, *Russ. J. Bioorg. Chem.*, 2004, **30**, 15.

47. D.S. Shah, T. Sakthivel, I. Toth, A.T. Florence and A.F. Wilderspin, *Int. J. Pharm.*, 2000, **208**, 41.

48. F. Tajarobi, M. El Sayed, B.D. Rege, J.E. Polli and H. Ghandehari, *Int. J. Pharm.*, 2001, **215**, 41.

49. M. El-Sayed, M. Ginski, C.A. Rhodes and H. Ghandehari, *J. Controlled Release*, 2002, **81**, 355.
50. R. Jevprasesphant, J. Penny, D. Attwood, N.B. McKeown and A. D'Emmanuele, *Pharm. Res.*, 2003, **20**, 1543.
51. G.T. Knipp, N.F.H. Ho, C.L. Barsuhn and R.R. Borchardt, *J. Pharm. Sci.*, 1997, **86**, 1105.
52. A. D'Emmanuele, R. Jevprasesphant, J. Penny and D. Attwood, *J. Controlled Release*, 2004, **95**, 447.
53. R. Wiwattanapatapee, B. Carreño-Gômez, N. Malik and R. Duncan, *Pharm. Res.*, 2000, **17**, 991.
54. M. El-Sayed, M.F. Kiani, M.D. Naimark, A.H. Hikal and H. Ghandahari, *Pharm. Res.*, 2001, **18**, 23.
55. E. Uzgiris, *Invest. Radiol.*, 2004, **39**, 131.
56. H. Kobayashi, N. Sato, S. Kawamoto, T. Saga, A. Hiraga, T.L. Haque, T. Ishimori, J. Konishi, K. Togashi and M.W. Brechbiel, *Bioconjugate Chem.*, 2001, **12**, 100.
57. H. Kobayashi, S. Kawamoto, T. Saga, N. Sato, A. Hiraga, T. Ishimori, J. Konishi, K. Togashi and M.W. Brechbiel, *Magn. Res. Med.*, 2001, **46**, 781.
58. D.V. Sakharov, A.F.H. Jie, D.V. Filippov, M.E.A. Bekkers, J.H. van Boom and D.C. Rijken, *FEBS Lett.*, 2003, **537**, 6.
59. D.V. Sakharov, A.F. Jie, M.E. Bekkers, J.J. Emeis and D.C. Rijken, *Arterioscler. Thromb. Vasc. Biol.*, 2001, **21**, 943.
60. L.D. Margerum, B.K. Campion, M. Koo, N. Shargill, J.J. Lai, A. Marumoto and P.C. Sontum, *J. Alloys Compd.*, 1997, **249**, 185.
61. A. Santana, S. Hyslop, E. Antunes, M. Mariano, Y.S. Bakle and G. de Nucci, *Agents Actions*, 1993, **39**, 104.
62. T. Sakthivel, I. Toth and A.T. Florence, *Int. J. Pharm.*, 1999, **183**, 51.
63. S.S. Nigavakar, L.Y. Sung, M. Llanes, A. El-Jawahri, T.S. Lawrence, C.W. Becker, L. Balogh and M.K. Khan, *Pharm. Res.*, 2004, **21**, 476.
64. C. Plank, K. Mechtler, F.C. Szoka and E. Wagner, *Human Gene Ther.*, 1996, **7**, 1437.
65. S. Shaunak, S. Thomas, E. Gianasi, A. Godwin, E. Jonnes, I. Teo, K. Mireskandari, P. Luthert, R. Duncan, S. Patterson, P. Khaw and S. Brocchini, *Nature Biotechnol.*, 2004, **22**, 977.

CHAPTER 3

Dendrimers as Drug Delivery Devices

3.1 Introduction

Drug delivery is an important aspect in the formulation of a drug because the proper choice of delivery system can control the bioavailability, concentration profile and undesirable side effects (by targeted delivery). Drug delivery is a large field covering on the one side, the problems of getting a drug into the patient in the simplest possible way and on the other, how to ensure that a compound for chemotherapy gets only to the cancer cells targeted.

The simplest way in terms of being user friendly is of course if the drug can be taken orally, but this requires that the drug is stable to the conditions (pH, enzymatic activity, epithelial permeability). Peptide- or oligonucleotide-based drugs (insulin is a classic example of a peptide drug) will normally be degraded by digestive enzymes. So in such cases, a suitable drug-delivery system should protect the drug against degradation as well as ensure that the drug achieves the proper permeability properties *e.g.* the ability to pass into the bloodstream. Once inside does not mean that the problem is solved, because there are a number of natural barriers that may have to be bypassed. An example, which relates to the topic of this chapter (Section 3.8), is gene therapy. In gene therapy, the DNA needs to be transported into the nucleus of the target cell, and in order to get there, it has to pass the outer membrane of the cell, has to survive the nucleases in the cell and should have the ability to get inside the nucleus. This is not possible without a combined transportation and protection system which may be a virus or even safer a dendrimer.

Controlled release is important in cases where it is impossible or impractical to use normal delivery systems. One example is the use of degradable polymers to encapsulate a chemotherapeutic drug, where a pad of drug-containing polymer is left *e.g.* in the cavity formed after surgical removal of a tumour slowly releasing the drug and thereby killing any residual cancer cells.[1] Another example is the deposition of contraceptive steroids in fatty tissues giving slow release into the bloodstream and long-term protection against pregnancy.

Large dendrimers and polymers both belong to the class of macromolecules, but they differ in the sense that polymers (including hyperbranched polymers) are always mixtures of compounds structurally closely related with different molecular weight, where a dendrimer is a single well-defined compound *i.e.* monodisperse. Drug delivery utilising polymers were originally introduced by Ringsdorf,[2,3] Kopeček[4] and Duncan.[5] The main problem associated with using polymers for drug delivery is the broad molecular weight distribution often found in polymers, which in terms can lead to irreproducible pharmacokinetic behaviour.

Dendrimers and polymers have some properties in common, when considering drug delivery. Cellular uptake of dendrimers (and polymers) can take place by endocytosis in cells and can thus bring drugs "bound" to the dendrimer into the cell. Both classes of compounds are also showing the so-called enhanced vascular permeability and retention (EPR) effect, which makes dendrimers and polymers attractive for targeting solid tumours.[6] Tumour vasculature has an increased permeability and limited lymphatic drainage, so this will over time lead to an accumulation of the macromolecules in the tumour. However, the remaining of this chapter deals with dendrimers for drug delivery,[7] and for further information on polymers for drug delivery there is a number of reviews available.[8–14]

Toxicity and biodistribution of dendrimers are important in medical use of any drug, and for dendrimers this is highly dependent on the actual structure of the dendrimer in question.[15,16]

The concept of dendritic growth leads to the formation of several compartments in the dendritic molecule, and the size of the compartments depends on the actual substructural units of the dendrimer. The dendritic structure can roughly be divided into three parts: (1) a multivalent surface, with a high number of potentially reactive sites, (2) an intermediate region ("the outer shell") having a well-defined microenvironment giving some protection from the outside by the dendrimer surface, and (3) a core, in which higher generation dendrimers are shielded from the surroundings, creating a microenvironment surrounded by the dendritic branches.

Drug delivery with dendrimers can essentially take place by two distinct types of mechanisms: (1) by *in vivo* degradation of a drug–dendrimer conjugate, where the drug is covalently bound to the dendrimer, or (2) by utilising host–guest chemistry where the drug is present as a guest in the dendrimer, and is released due to changes in the physical environment such as pH, temperature or simply released by diffusion out of the dendrimer.

An *in vivo* degradation is of course dependent on the presence of suitable enzymes or on an environment capable of degrading the covalent bonds in question, where the approach based on host–guest chemistry can be independent of external factors.

Two other important issues in drug delivery are targeted delivery to a specific type of cell or controlled release from a depot, which may be present in circulation or imbedded in some suitable tissue. Host–guest based systems are treated first in the following sections, since they are the most studied dendrimer-based drug delivery systems.

The host–guest binding can either take place in the cavities of the dendrimer core ("endo-receptor"), or at the multivalent surface or outer shell of the dendrimer ("exo-receptor").

3.2 Dendrimer Hosts

The important questions when dealing with host–guest chemistry in dendrimers are how to prove the existence of the host–guest complex (and ideally the stoichiometry) and how to show where the guest is located in the dendrimer and the binding mode between the dendrimer and the guest. Some representative techniques that have been proven useful are: the use of dyes or fluorescent compounds as guest molecules, using compounds, where the optical properties of the probes are sensitive to the surroundings (pH, polarity, *etc.*). Nuclear magnetic resonance (NMR) is a highly useful technique for studying dendrimer–drug complexes. Simple 1D experiments, typically[1] H-NMR, can be used for studying the formation of the host–guest complexes to provide information on the stoichiometry of the complex and binding modes (hydrogen bonding, ion pairing, *etc.*), while 2D experiments such as NOESY can give information on the spatial nature of the binding site. The techniques developed for studying protein structure by an NMR have recently also been successfully applied on dendrimers giving a structural insight into dendrimers in solution.[17,18]

3.2.1 Dendrimer Hosts: Non-Specific Interactions with the Dendrimer Core

Micelles are well-known structures in biological systems. They are formed from amphiphilic molecules having both a polar and an apolar part, a classical example being solutions of soap in water. Drug delivery by micellar systems, where the drug is encapsulated inside the micelle, is well known. But when the concentration of the micelles gets below the critical micellar concentration (CMC), the system becomes thermodynamically unstable and drops the payload.

Dendrimers with an apolar interior structure and a polar surface can be viewed as unimolecular micelles, but due to their covalent nature they will not have a CMC. The pioneering work on unimolecular micelles was made by Newkome and co-workers,[19–22] who synthesised and studied the class of compounds called micellanoic acids (Figure 3.1). A special feature of these compounds is that the dendrimer is completely based on a carbon–carbon bonded skeleton, which makes them rather unique among the different families of dendrimers. The encapsulation of hydrophobic molecules (*e.g.* the dyes phenol blue and pinacyanol chloride and simple aromatic molecules such as naphthalene and diphenylhexatriene) in these dendrimers was proven by UV–VIS and fluorescence spectroscopy. Micellanoic acids have hydrophilic carboxylic acid groups at the surface ensuring solubility in water and a hydrophobic interior, and these compounds can potentially be used as hosts for solubilising hydrophobic drugs in water.

A system, where the uptake/release is purely controlled by pH of the surroundings is the fatty acid-derivatised poly(propylene imine) (PPI)-dendrimers. These compounds behave as unimolecular-inverted micelles and were shown to encapsulate guest molecules such as rose bengal,[23] and a series of anionic dyes could be extracted from water into organic solvents.[24] The extraction process is pH-dependent in the sense that some degree of protonation of the interior amino groups is necessary before extraction of the anion takes place. Furthermore, a pH dependence was also found for extraction of different dyes, and the utility of this was demonstrated by the separation of rose bengal from fluorescein simply by controlling pH in the aqueous phase.

The number of dye molecules present in the dendrimers depends on the size of the dendrimer and the guest molecule. In the case of a G5-PPI dendrimer with fluorescein,

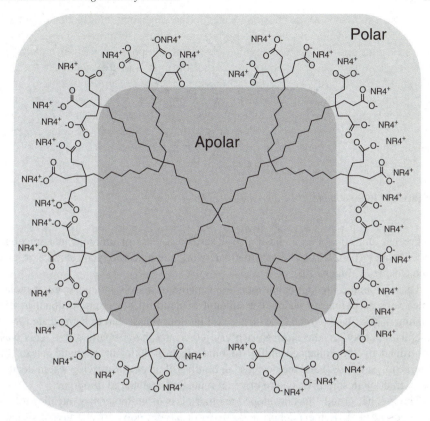

Figure 3.1 *A micellanoic acid for binding of hydrophobic guests such as naphthalene and phenol blue*

only 1–2 dye molecules were encapsulated per dendrimer, whereas in the case of rose bengal up to 70 dye molecules could be encapsulated. The stoichiometry of the complexes was determined by UV–VIS spectroscopy. The development of such systems may potentially be useful for depositing polar drugs in fatty tissues.

Recently, the surface groups were changed from simple fatty acid amides to a PEG-based structure enhancing the solubility both in water and organic solvents as well as the ability to complex anionic dyes (Figure 3.2).[25] The amides at the surface of the dendrimer form a hydrogen-bonded shell around the dendrimer and the hydrogen bonds are protected from the surrounding water by the phenyl groups. The entire molecule becomes soluble due to the tetraethyleneglycol methyl ether at the periphery of the benzene rings. The binding of Tetrachlorofluorescein and bengal rose was investigated by UV–VIS, and the structure of the dendrimers in solution as well as the location of the guests in the complexes were investigated by small angle X-ray scattering (SAXS). These experiments proved that the guests were in fact located inside the dendrimers in these complexes and not simply bound to the surface.

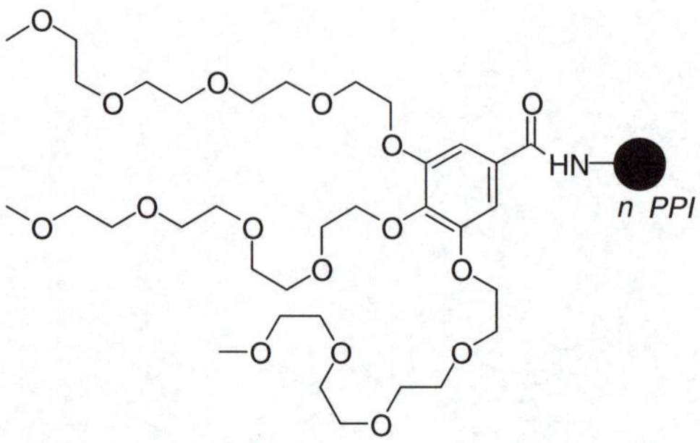

n = 4, 8, 16, 32, 64

Figure 3.2 *Water-soluble PPI-dendrimer-based hosts*

Fréchet and co-workers[26] have also described a class of unimolecular micelles based on poly(benzyl ether) dendrimeric network having carboxylate surface groups (see Chapter 1, Figure 1.6).

These aromatic Frechét-dendrimers are capable of dissolving apolar guest molecules such as pyrene in water. The amount of dendrimer was proportional to the amount of dissolved pyrene. The host–guest binding is assumed to be mediated $\pi-\pi$ interactions between the electron-rich aryl ether and the aromatic guest. This was confirmed by the enhanced ability to bind electron-deficient aromatic guests (π interaction stabilised) and the decrease in binding of an electron-rich guest molecule (π interaction destabilised due to electron repulsion) compared to pyrene.

The Fréchet group[27] has developed another class of unimolecular micelles based on a poly(arylalkyl ether) dendrimeric network having polyethylene glycol surface groups. The reasons for using PEG tails are four-fold: first of all it is a simple way of increasing the hydrodynamic ratio of the molecule, which is important for controlling the circulation time of the dendrimer. The action of the kidneys can simply be described as a filter, where excretion of a compound from the plasma is dependent on the actual size of the molecule (the hydrodynamic ratio) (see also Chapter 2). The interior of the dendrimer is highly apolar, and was found to be suitable for carrying the antiinflammatory drug indomethacin with a loading of 11%. Controlled release was shown in *in vitro* by dialysis experiments.

As the inverted micelle, PAMAM-dendrimers with an apolar core (1,12-diaminododecane) bind apolar compounds such as the dye Nile Red. Generation dependence was found for the binding of Nile Red in water with a decrease in binding on going to higher (>G4) generation dendrimers (Figure 3.3). Upon increasing the generation of the dendrimer, the number of carboxylic acid groups at the surface increases, and this leads to an increasingly polar surface. The dye being highly apolar is thus prevented from accessing the core.[28]

The use of an internally branched PAMAM-dendrimer for carrying the anticancer drug fluorouracil has been studied by Jain and co-workers,[29] who also investigated the effect of substitution of the periphery with PEG groups (Figure 3.3). They found a significant difference between the amino terminated and the PEGylated dendrimers. PEGylation was found to reduce the haemolytic activity, increase the loading capacity

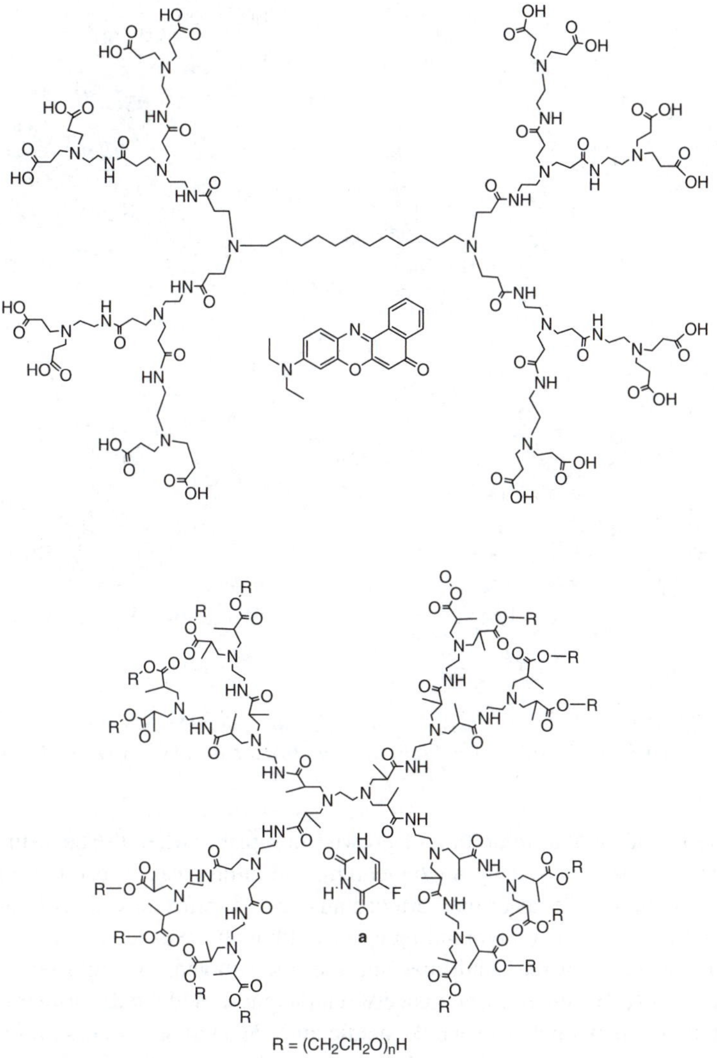

Figure 3.3 *Two PAMAM systems utilised as hosts for drugs. Top: G2.5 PAMAM dendrimer with a 1,12-diaminododecane core shown with a guest molecule Nile red. Bottom: G2.5 Me-PAMAM dendrimer with PEG tails investigated by Jain's group[29] shown with fluorouracil (a) as guest*

as well as reducing the drug-release rate in albino rats.[16] The type of dendrimer that Jain and co-workers have described is apparently the first example of a dendrimer of the PAMAM family based on methacrylamides as structural elements.

Hydroxy-terminated PAMAM dendrimers (Figure 3.4), which are water-soluble dendrimers, are capable of binding acidic antifungal or antibacterial aromatic compounds.[30] Flow microcalorimetry was used to observe the effect upon the growth of yeast cells. By this technique, the heat evolution from the biological processes was

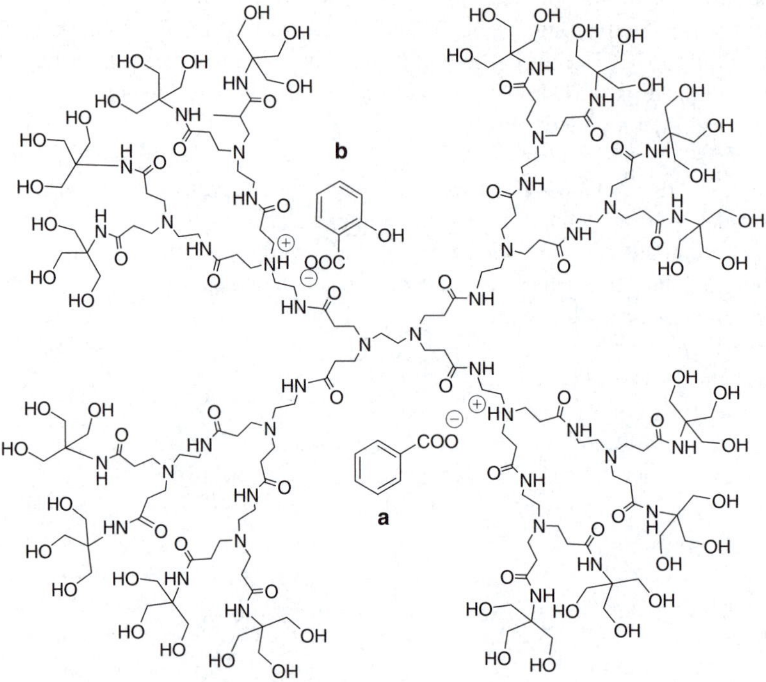

Figure 3.4 *G-2.5 tris(hydroxymethyl)methylamide-terminated PAMAM-dendrimer shown with benzoic acid guests (a) and salicylic acid (b) as studied by Twyman and co-workers*[31]

monitored showing that inhibition of growth took place, when the dendrimer–drug conjugate was present in the growth medium, and killing was observed after 24 h if the concentration of the dendrimer–drug conjugate was sufficiently high (2 mg mL^{-1}). The dendrimer itself had no inhibiting effect. Although the exact nature of the complex binding between the dendrimer and the guest molecules could not be elucidated, *e.g.* by NMR, the interactions between the guests and the dendrimer may rely on acid–base interactions between the acidic guest and tertiary amines present at the focal points of the dendrimer. The ionic nature of the complex bond is supported by the fact that the apolar compound Tioconazole did not form complexes with the dendrimer.

Triazine-based dendrimers have been prepared for drug delivery.[32–34] The dendrimers were studied as solubilising agents for the drugs indomethacin, methotrexate and 10-Hydroxycamptothecin. The ability to solubilise pyrene was tested among the triazine dendrimer and PAMAM, PPI and polyphenyl ether dendrimers that are common types of dendrimers. The latter has been suggested as a slow-release system for indomethacin.[27] The triazine dendrimer was comparable with the polyphenyl ether dendrimer with respect to solubilisation of pyrene in water. This is unexpected, since triazine is a rather polar heterocyclic compound and therefore quite different in properties than an apolar polyphenyl ether, however, attractive interactions between the apolar pyrene and the triazines may be provided by π–π bonding.

The drug-delivering ability of the triazine dendrimer was subsequently tested *in vitro* as carrier for the drug 1,1-bis(3-indolyl)-1-(*p*-trifluormethylphenyl)methane in a cell culture (MCF-7 cells). The conjugate was found to be equally efficient in a solution of the drug in DMSO, which is a commonly known solvent with a high ability to transport compounds across membranes.

The *in vivo* toxicity on liver and kidneys in mice was also investigated, and no toxic effects were observed in single doses delivered intraperitoneally up to 10 mg kg^{-1}.[32–34]

Transdermal drug delivery is of general interest, because it is an alternative to oral administration allowing direct delivery of the drug through the skin into the tissue. Transdermal delivery of indomethacine has been shown to benefit from a composition containing PAMAM-dendrimers.[35,36] Indomethacine was used as a model drug, and the highest transportation rates through skin were observed with PAMAM-dendrimers G4-PAMAM and G4-PAMAM-OH.

PAMAM-dendrimers have also been investigated as ophthalmic vehicles for delivery of compounds such as pilocarpine, tropicamide (Figure 3.5) and fluorescein, all drugs routinely used in ophthalmology for diagnostics.[37] This was done with an *in vivo* model based on New Zealand albino rabbits. One of the problems with ocular drug delivery was to increase the residence time in the eye. The PAMAM-dendrimers tested had OH, NH$_2$ or COOH as end groups, and two differently sized dendrimers (a G3 and G5) were tested for each type of end group. The dendrimers having -OH or -COOH as end groups were found to provide better bioavailability and residence time in the eye than –NH$_2$ terminated PAMAMs. The residence time was found to depend on the size and molecular weight of the dendrimers, and this could be a future drug-delivery system for the treatment of ocular diseases.

A generation 3.5 PAMAM-dendrimer[38] conjugated to the well-known anticancer drug *cis*platin acts as macromolecular carriers for platinum. The dendrimer-platinate gives a slower release of the platinum because the interior of the dendrimer can act as a ligand for platinum, and complexes of Pt(II) are known to be kinetically very stable. The dendrimer-platinate also does show higher accumulation in solid tumours – and lower toxicity – compared to *cis*platin.[39] This is probably due to the EPR effect, which is an apparent feature of the dendrimers. The increased permeability and limited lymphatic drainage lead to an accumulation of the dendrimer-platinate in the tumour tissue, and consequently a larger local concentration of Pt than possible with an unbound *cis*platin.

A new family of dendrimers, designed with drug delivery in mind is the so-called "bow tie" dendrimers[6] (Figure 3.6). These are asymmetric dendrimers that consist of two dendritic wedges having the focal point in common. These structures are

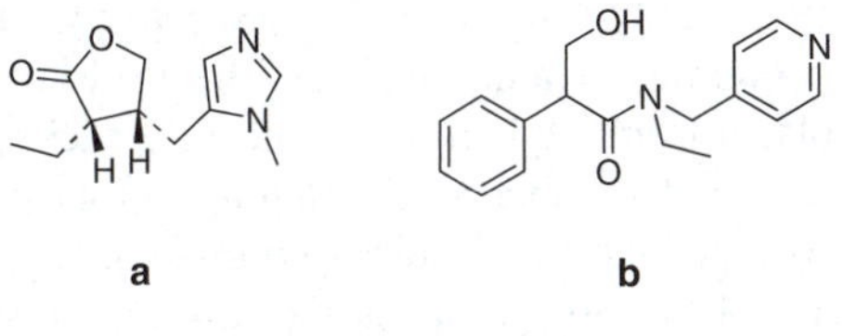

Figure 3.5 *Pilocarpine (a) and tropicamide (b)*

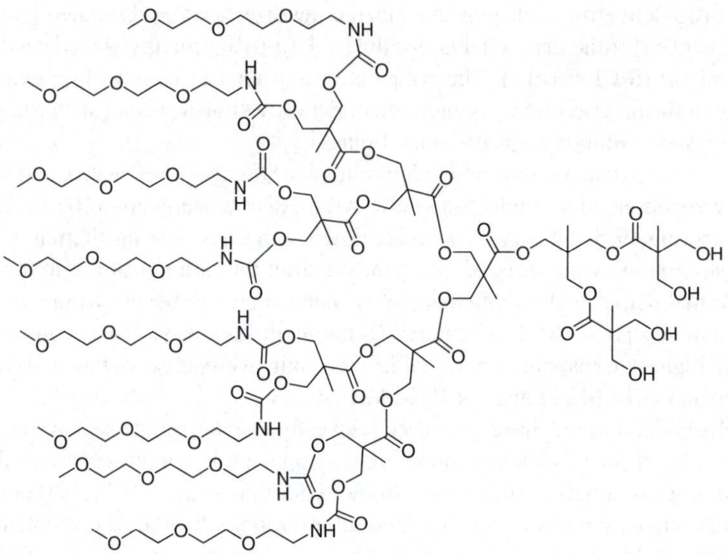

Figure 3.6 *A small Fréchet "bow-tie dendrimer"*

essentially non-toxic towards MDA-MB-231 cancer cells. A radioactive tracer (^{125}I-labelled tyramine) was conjugated to these dendrimers via a carbamate linker. They are biodegradable under physiological conditions due to the structure containing ester- and carbamate-groups. The biodistribution has been investigated, and high levels of tumour accumulation were found in mice bearing subcutaneous B16F10 melanoma. The half-life in circulation could be controlled by the length of the PEG chains on the side of the bow tie.

3.2.2 Dendritic Boxes or Topological Trapping of Guests

The systems described in the previous section are reversible in the sense that the guest molecule or drug can enter or leave the dendrimer freely depending on the actual external stimuli. The "dendritic box",[40–46] was the first example of dendrimer-based host–guest complex, where the guest molecules were physically trapped inside the dendrimer. They were not covalently bound, but due to the structure of the dendrimer, the guest molecules could only be released by degrading the shell of the dendritic host. The original dendritic box was based on a G5-PPI dendrimer modified at the surface with numerous Boc-protected phenylalanines. In this way, the outer shell was made denser due to the sterically demanding Boc-protective groups. Guest molecules of different size, present during the modification of the dendrimer, were encapsulated in the interior and isolated from the bulk by the densely packed Boc-phenylalanine surface. The dendrimer could simultaneously bind up to four large guest molecules (rose bengal) and 8–10 small guest molecules (*p*-nitro-benzoic acid). Upon selective acidolysis (formic acid) of the Boc-groups at the surface, the surface shell became more open and the small guest molecules were allowed to leak from the dendrimer, whereas the large guest molecules remained trapped in the core. The large guest molecules could subsequently be released from the dendrimer by acidolysis of the amide bonds creating the unmodified dendrimer with a more open surface structure. In the dendritic box, the interactions between the host and the guest molecule were not tailored to be specific, but more governed by the molecular size of the guest molecule, and the physical size of the cavities in the host. The concept of the "dendritic box" could be envisioned for using in stepwise delivery of different compounds differing sufficiently in size. The use of natural building blocks such as amino acids or peptides on the dendrimer surface may open to enzymatic degradation of the dendrimer surface. However, the "loading of the gun" cannot be performed on the modified dendrimer, but has to take place during the synthesis of the box.

Metallic silver has long been known to have antimicrobial activity, and dendrimer-encapsulated silver nanoparticles in hydroxy-terminated PAMAM-dendrimers have been shown to slowly release silver, and thereby having antimicrobial activity against various gram-positive bacteria.[47] This system, belonging to the class of dendrimer-encapsulated metal nanoparticles, is made in two steps from the PAMAM-dendrimer taking advantage of the ability of the interior amino groups of the dendrimer to form complexes with metal ions such as Ag^+. After loading of the dendrimer with Ag^+, a reducing agent capable of reducing Ag^+ to Ag is added, and the "silver-bullet" is cast inside the dendrimer. The dendrimer slowly releases antibacterial Ag^+ ions.

3.2.3 Dendrimer Hosts: Specific Interactions with the Dendrimer Core

The preceeding dendrimers did not have specific binding sites, and thus may in principle exchange their guests with other guests from the surroundings. Dendrimers with specifically tailored binding sites could be useful for carrying specific drugs without having to worry about the dendrimer acting as a non-specific sponge that might create problems by picking up various biomolecules on its way. Diederich's group has created a family of water-soluble dendrimers with specific binding sites, the "dendrophanes". The dendrophanes are centered on a "cyclophane" core, and can bind aromatic compounds, presumably via π–π interactions (Figure 3.7). These dendritic structures have been shown to be excellent carriers of steroids,[48] and may potentially be used for a controlled release. The stability of the complexes was not affected by the generation of the dendrimers, and the binding site was found to favour non-polar compared to polar steroids.

In order to be able to bind more polar bioactive compounds to the core of a dendrimer, Diederich's group designed another class of specialised dendritic water-soluble hosts, the so-called "dendroclefts".[49,50] These dendrimers were centered around an optically active 9,9'-spirobi[9*H*-fluorene] core and showed a marked diastereoselectivity towards recognition of octyl β-D-glucoside over octyl α-D-glucoside.[1] The H-NMR analysis performed on the host–guest complexes showed that hydrogen bonding between the pyridine carboxamide moieties in the core and the oxygen atoms in the carbohydrate guest were the major contributions to the host–guest interaction. The diastereoselectivity was found to increase with increasing dendrimer generation, probably due to increased hydrogen bonding between the bound carbohydrate guest

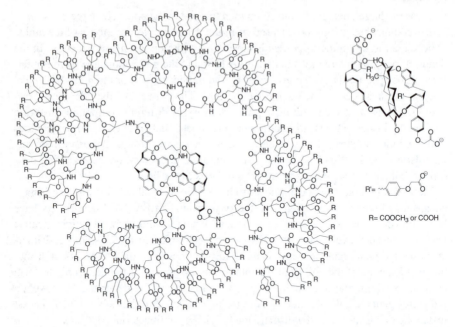

Figure 3.7 *Dendrophanes which are dendrimers born with a specific binding site for steroids*

and the alkyl ether oxygen atoms of the dendritic wedges.[49] The high stereo selectivity found in the selection of guest molecules of the dendroclefts make these interesting as systems mimicking the highly specific intake of substrates in enzymes or other carbohydrate carriers found in nature. A dendrimer with internal binding motifs for molecular recognition by hydrogen bonding was prepared for specific binding of imide-containing drugs.[51] The "Newkome-system" could bind complementary guests such as Glutarimide, AZT or barbituric acid. Barbituric acid was found to bind to the host with a $K_a = 70$ M^{-1} in CDCl$_3$ as determined by[1] H-NMR (Figure 3.8).

Such a system will however need to be optimised for obtaining higher association constants, before uses for drug delivery can be envisioned.

3.2.4   Dendrimer Hosts: Non-Polar Interactions with the Dendrimer Surface Group

Cyclodextrins (CD) are a class of complex cyclic carbohydrates, which have been extensively studied in supramolecular chemistry due to the combination of availability and physical size. They have also found use in biotechnology[52] and pharmaceutics.[53,54] CDs, which are prepared by bacteria, can be described as "barrels made of sugar" that have a hydrophilic exterior and a hydrophobic interior. These were first utilised in connection with dendrimers by Kaifer and co-workers,[55] who described a system based on a cobaltocenium substituted PPI-dendrimer, where the multivalent binding of β-cyclodextrins (β-CD) could be controlled by reduction of the cobaltocenium units (charge $+1$) to cobaltocene (charge 0). The change in charge results

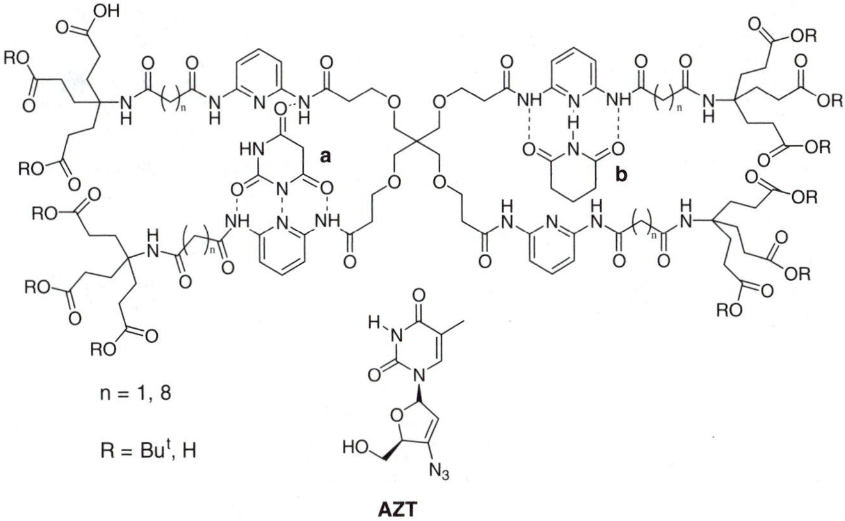

Figure 3.8 *G1 Newkome system shown with barbituric acid (left side a) and glutarimide (right side b) as guests*

in a drastic change of polarity from being highly polar in the +1 state to highly apolar in the reduced state facilitating binding in the apolar interior of the CD. This principle of electroactive drug release could be imagined to be triggered by enzymatic oxidation, for example, cytochrome P450 in a biological system, and thereby releasing the encapsulated drug (Figure 3.9).

Studies interestingly also show that adamantyl urea substituted dendrimers could be solubilised in water by the formation of a host–guest complex with β-CD. However in contrast to the Kaifer system, these are not electroactive, and the dendrimer is merely a passive carrier of β-CD (Figure 3.9, bottom).[56]

These two examples show that it is possible to solubilise highly apolar dendrimers by a formation of host–guest complexes with β-CD. The increased solubility and stability of the hydrophobic binding between the hydrophobic phase of β-CD and adamantyl upon binding the CD to the dendrimer may be a driving force for the complex formation. These complexes showed good water solubility and thus may potentially be useful as CD vehicles under physiological conditions.

3.2.5 Dendrimer Hosts: Polar Interactions with the Dendrimer Surface Groups

The majority of dendrimers has polar groups at their surfaces, typically NH$_2$- or COOH-groups, create the option of using the dendrimer as either an acid or a base for salt-formation with a drug. Kannan's group[57] showed that salt formation took place as expected between an acidic drug such as ibuprofen and PAMAM-dendrimers of generation 3 or 4 with NH$_2$-groups at the surface. An *in vitro* release from the drug–dendrimer complex was much slower when compared to ibuprofen,

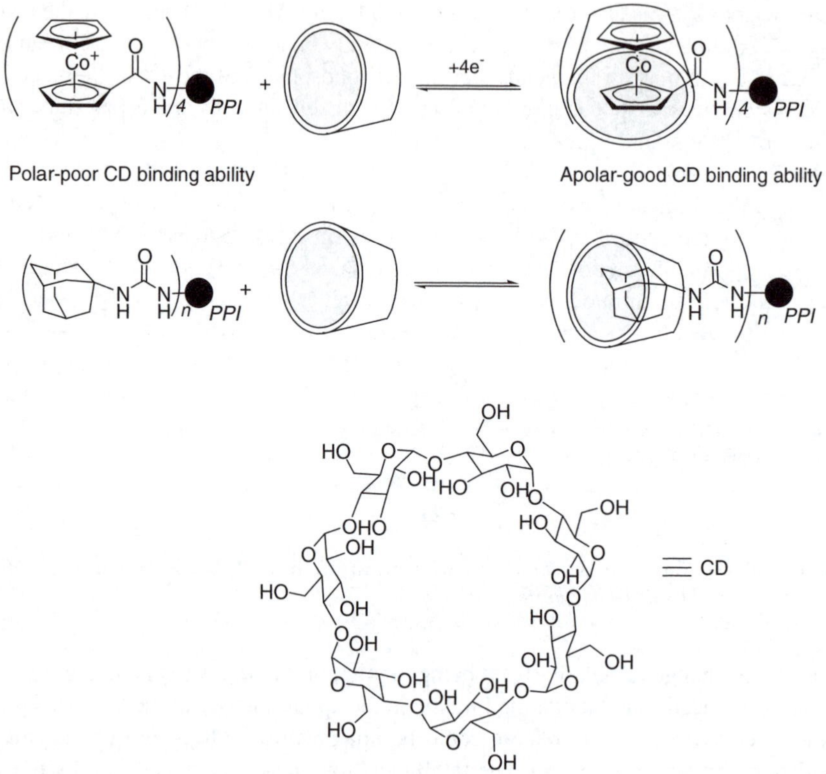

Polar-poor CD binding ability   Apolar-good CD binding ability

≡ CD

Figure 3.9 *Two types of CD hosts, top: Kaifer's reversible guest–host system, bottom: PPI dendrimers having adamantyl surface groups can act as hosts for CD*

and the complexed drug entered A549 cells much faster than the drug itself. This is interesting since this is almost the simplest type of system imaginable (a salt between a polycation and an anion). However such a polycation–anion system will not be specific in the sense that any anion regardless of its origin could bind to the dendrimer, and this could give complications if biologically relevant anions were removed locally. A possible solution to this problem would be to have a system where additional interactions could provide a specificity for certain types of compounds.

The structure of the end groups of dendrimers such as the PPI-dendrimers consists of bis(3-aminopropyl)amine, so the dendrimer can be viewed as having a number of pincers presented towards the exterior. Pincer structures are often used in host–guest chemistry for construction of hosts. Modification of the surface of PPI-dendrimers with amides, ureas or thioureas leads to systems, where strong hydrogen bonding occurs between the amide NHs at the surface. This led to the development of urea-substituted PPI-dendrimers as hosts for host–guest complexation, where the binding takes place at the surface of the dendrimer.

Several types of guest molecules are possible, and one example is the binding of oxyanions to urea-modified PPI-dendrimers shown by Vögtle and co-workers.[58] They showed binding of the biologically important phosphates ADP, AMP and ATP.

Some selectivity was also observed for the binding of ADP, AMP and ATP, the exact mode of binding was not investigated, but may rely on ionic interactions between the basic tertiary amines in the dendrimer and the phosphates in the "AXP" molecules. The system was also capable of binding pertechnetate, which is an important radioactive compound used in diagnostics.[59]

Meijer's group[60,61] utilised urea- and thiourea-functionalised PPI-dendrimers for binding guest molecules containing a urea-glycine "tail" unit. The guest molecules interact with the dendritic host by multiple urea (guest)-(thio) urea (host) hydrogen bonds and ionic interactions between the glycine carboxylic acid and the dendrimer outer shell tertiary amino groups. It has been suggested that the acid–base reaction between the dendrimer and guest with subsequent Coulombic attractions, pulls the guest into the dendrimer, whereas the hydrogen bonding keeps the guest bound to the host. By the intake of urea guests, the outer shell becomes increasingly crowded and dense, hence this host–guest motif could provide a non-covalent example of a dendritic box (Figure 3.10).

As the urea glycine tail is highly similar to the *C*-terminus of a peptide, it has been investigated whether the dendrimer could act as a host or carrier for peptides, directing this "click in" motif towards biological applications. This host–guest motif may be useful as a pH sensitive drug-delivery system for peptide drugs, otherwise sensitive to proteolytic degradation. The peptide is released upon lowering the pH, as a result of protonation of the carboxylic acid moiety, strongly diminishing the association between the peptide and the dendrimer. It was found that the urea-and thiourea-modified dendrimers were capable of binding different peptides, regardless of the bulkiness of the side chains, and that the peptides could be released from the dendrimer under mild acidic conditions.[62] As the dendrimer was able to bind peptides with different side chain motifs, this would introduce the possibility of using the

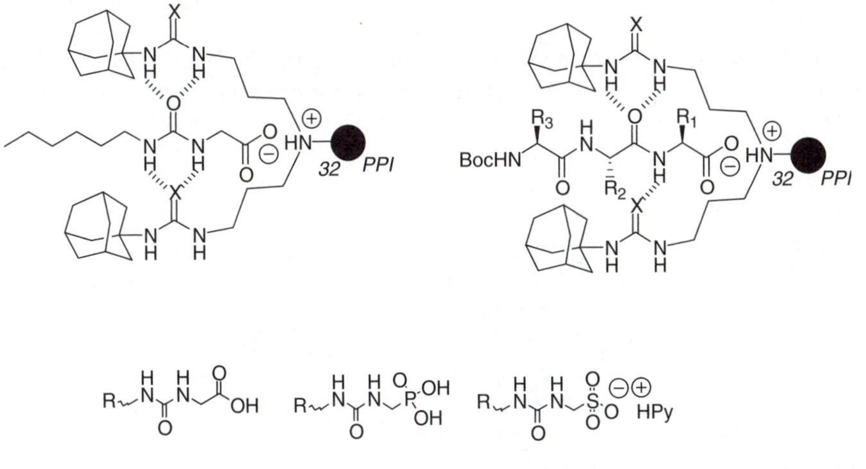

Figure 3.10 *The 'click-in' design. Left: intake of urea-containing guest molecules in urea or thiourea hosts.[61] Right: intake of N-Boc-protected peptides,[62] X: O, S a: Ureido carboxylic acid guest. b: Ureido phosphonic acid guest. c: Ureido sulfonate guest. The order of binding is: a < b < c*

dendrimer as host ("bus") for several different peptides ("passengers") simultaneously. A pH-dependent drug-delivery system is interesting from a biological point of view as cellular uptake by phagocytosis gives rise to significant changes in the pH.

The guest motif has been further improved, so that a series of guest motifs covering a wide range of association constants are now available (Figure 3.10).[63]

Another system based on hydrogen bonding in Hamilton-receptor functionalised PPI-dendrimers was described recently by Vögtle and co-workers[64] (Figure 3.11). The association constants determined by[1] H-NMR were between 10^3 and 10^5 M^{-1} for different barbiturate guests in a model system. The binding motif of the Hamilton-receptor has some structural similarity to the pyrimidine bases from DNA and RNA, and thus it might be possible to develop a oligonucleotide carrier based on this type of system.

3.3   Covalently Bound Drug-Dendrimer Conjugates

The systems described in the previous sections were all based on non-covalent binding of the drug to the dendrimer and release could occur by changes in the physical environment with the exception of "dendritic box" systems. However, the formation of host–guest complexes may not always be possible or desirable with a given drug and a dendrimer. A general problem with systems for drug delivery is how to achieve a sufficiently large payload. This makes dendrimers attractive, because they have a large number of surface groups, where the drug can be covalently attached, thereby transforming the drug into a pro-drug.

Dendrimers based on a 1,4,7,10-tetraazacyclododecane core having primary amines at its surface have been partially modified by acylation with 1-bromoacetyl-5-fluorouracil to form a labile imide linkage. The binding to a dendritic vehicle may reduce the toxicity of 5-fluorouracil and allows a slow release. The imide linkage was shown to hydrolyse under physiological conditions releasing 5-fluorouracil *in vitro*[65] (Figure 3.12).

In addition to the specifically tailored dendrimer scaffolds, commercially available PAMAM-dendrimers have been used as platforms for covalent attachment of drugs.

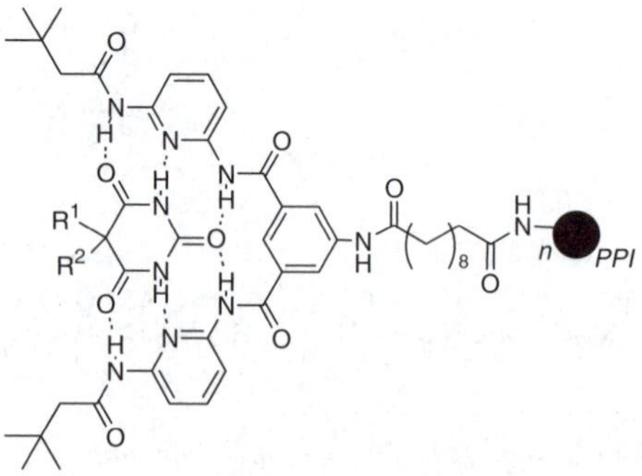

Figure 3.11 *PPI-dendrimer substituted with a Hamilton-receptor at the surface shown with a barbituric acid derivative in the binding site*

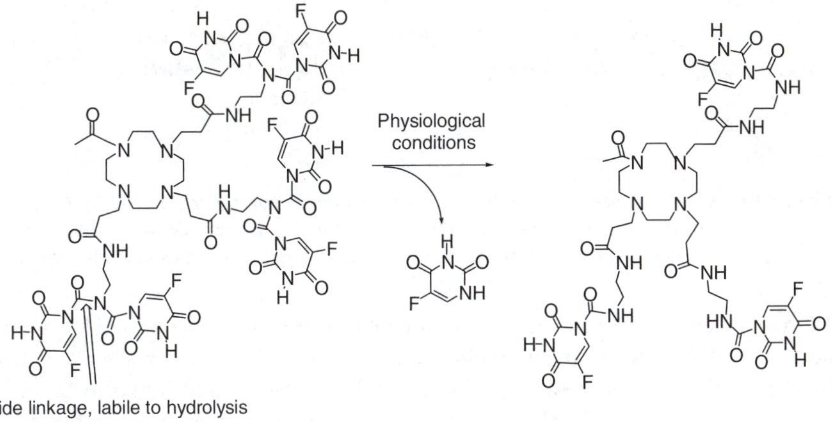

Imide linkage, labile to hydrolysis

Figure 3.12 *Structure of 1,4,7,10-tetraazacyclododecane cored dendrimer carrying 5-fluo-rouracil groups covalently bound to the dendrimer, the 5-fluorouracil can be released under physiological conditions by hydrolysis of the highly labile imide linkage*

Degenerative diseases of the colon such as Crohn's disease and ulcerative colitis can be treated with 5-aminosalicylic acid (ASA), which acts as a topical antiinflammatory drug. It is commonly used in the form of the pro-drug sulfalazine, which is reduced by bacteria in colon to ASA and sulfapyridine. While 5-aminosalicylic acid is not absorbed in the intestines, sulfapyridine is, and causes side effects such as hypersensitivity reactions in some patients. In order to solve this problem, PAMAM-dendrimers modified on the surface with ASA pro-drugs have been synthesised and tested[66](Figure 3.13). The conjugates were compared to sulfalazine in a rat model (male Wistar rats). Release was found to take place in the colon, but the commercial pro-drug had a significantly faster release of 5-aminosalicylic acid, due to a better enzymatic cleavage compared to the dendrimer system, so further development of the dendritic-delivery system is needed.

Steroids are another type of drugs, where it is desirable to control the delivery of the drug, because they have hormonal activity and may sometimes have serious side effects. Kannan and co-workers[67] have investigated the delivery of methyl-prednisolone with PAMAM-dendrimers. The construction of a suitable steroid–dendrimer conjugate presents some problems, because steroids are rather apolar and the few functional groups present (typically hydroxy groups) have low reactivity, which both have an effect in the synthesis of the conjugate as well as *in vivo*, where release can be slow. This is a natural consequence of the steroid skeleton, which is highly apolar causing the site of reaction to be less accessible for reaction with polar compounds such as water. Kannan and co-workers investigated two different PAMAM-dendrimers, a G 3.5 dendrimer with 32 COOH-groups and a G5-PAMAM-OH dendrimer with 64 OH-groups. The attachment of the steroid was done by direct ester formation to the G 3.5 dendrimer, and with a glutaric acid spacer in the case of the G5 hydroxy-terminated dendrimer. The small dendrimer was found to carry one methylpred-nisolone molecule per dendrimer, and the larger G4-dendrimer was found to carry

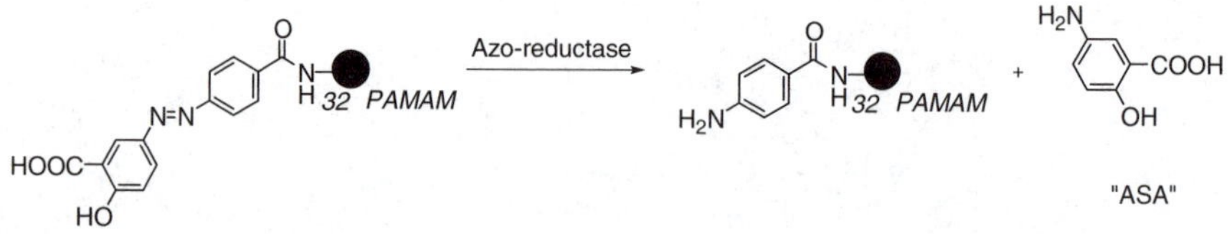

Figure 3.13 *G4-PAMAM-5-aminosalicylic acid conjugate investigated by Wiwattanapatapee and co-workers.*[66] *The linker shown in the figure was based on 4-aminobenzoic acid. The corresponding derivative from 4-aminohippuric acid was also investigated*

12 molecules per dendrimer. The conjugates were labelled with FITC and the uptake in A549 human lung epithelial cells was studied with fluorescence and confocal microscopy. The conjugates were rapidly taken up, and found to be mainly localised in the cytosol. The activity of the conjugates was comparable to free methylprednisolone.

Amino-terminated dendrimers are attractive for drug conjugates from a synthetic point of view, because of the well-developed methodology for formation of amides that has been created in the field of peptide synthesis. However, as previously mentioned, the presence of a large number of amino groups at the surface also leads to toxicity. A possible way of bypassing this problem would be to create hetero-functionalised dendrimers, with two types of groups bound to the surface: the drug, which should occupy part of the available sites, and another group, introduced to reduce the toxicity of the dendrimer after release of the drug. This strategy has been investigated by D'Emanuele and co-workers,[68] who did initial studies on the toxicity of amino-terminated PAMAM-dendrimers on Caco-2 cells. The dendrimers were either partially acylated with dodecanoic acid or partially alkylated with PEG groups. A marked reduction in cytotoxicity was observed, and this was ascribed to shielding of the charge from the remaining ammonium groups on the dendrimers by the tails added. Subsequently D'Emanuele and co-workers[69,70] showed that the modified dendrimers are internalised by endocytosis in Caco-2 cells. Given the possible application of dendrimers as carriers for chemotherapeutics as well as the fact that resistance towards chemotherapy tends to develop and that efflux mechanisms exist, where at least one seems to involve glycoprotein-P, the question of possible efflux of dendrimers was addressed. The glycoprotein-P efflux pump is known to be fairly broad in its range of substrates (see Chapter 2), and one of the known substrates is propanolol. This compound was conjugated to two G3-PAMAM-dendrimers, a pure amino terminated and a partially dodecanoylated dendrimer. It was found that the propanolol conjugates were taken in, but not pumped out of Caco-2 cells.[71] This means that drug–dendrimer conjugates with chemotherapeutics could have much more potential than being just circulating reservoirs for slow release of the drugs.

In order to obtain a specific cellular treatment, drug vehicles that direct the drug only to specific cell types can be designed. One example of such cell-specific dendritic drug vehicles is a dendrimer derivatised with folic acid (pteroyl-L-glutamic acid) (Figure 3.14). Folic acid is an important substrate for uptake in cells by the *folate receptor pathway*. As the folate receptor is overexpressed in cancer cells, these folic acid-derivatised dendrimers are taken up by cancer cells preferentially to normal

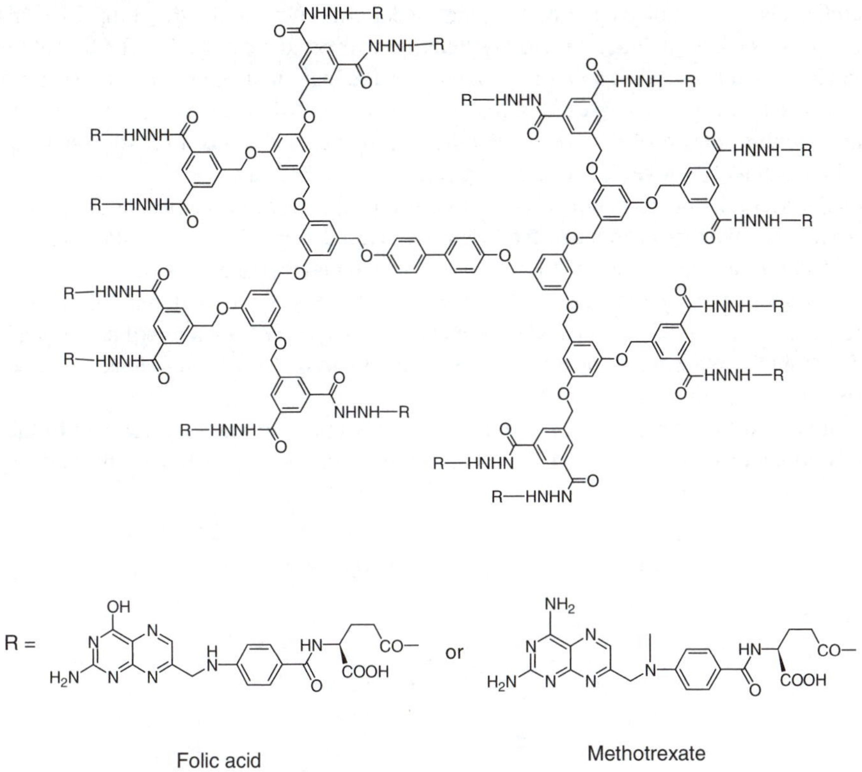

Figure 3.14 *Dendrimers carrying folic acid or methotrexate on its surface*

cells, making these dendrimers well suited for the cancer-specific drug delivery of cytotoxic substances.[72,73]

Recently Baker and co-workers[74] described a system based on conjugation of an antibody with a PAMAM-dendrimer. The antibody had specificity for a surface glycoprotein expressed by prostate cancer cells. They showed by fluorescence labelling that the antibody–dendrimer conjugate was taken up by the cancer cells, and it will be interesting to see this approach used for targeted delivery of chemotherapy.

3.3.1 Self-Immolative Systems

A self-immolative dendrimer is essentially a dendrimer that has been constructed in such a manner that it can "self-destruct" and decompose into small pieces upon biological stimuli. This is interesting because small molecules are easier excreted from the body via the kidneys. Another interesting aspect is that when the dendrimer disintegrates, it may release all the drugs bound to the dendrimer, and thereby creates a high local concentration. If delivery of a dendrimer–drug conjugate furthermore can be targeted at a specific type of cell, then systems having a "trigger mechanism" capable of delivering all of their contents could be interesting. The action of such a system would involve recognition and binding to the desired target, and a subsequent release of the conjugated drug molecules giving a high local concentration of the drug. The first reports of such immolative cascade systems were reported recently.[75–77] The first of the model systems described so far rely on chemical reduction of an aromatic nitro group. This reaction triggers a cascade of reactions, where the whole dendritic structure falls apart with a release of the conjugated drug, in this case Paclitaxel (Taxol®) together with small and relatively harmless molecules (Figure 3.15, top).

The second model system based on hydrolysis of an ester by penicillin G amidase, where the hydrolysis induces the complete degradation of the dendritic structure with simultaneous release of the surface groups, is phenyl acetic acid.[78]

An alternative system reported cleaves by activation with a catalytic antibody,[79] and has shown good properties as delivery system for the anticancer therapeuticals doxorubicin and camptothecin, in a cell-growth assay with Molt-3 leukemia cells (Figure 3.15, bottom).

It is clear that these model systems are merely a proof of the concept, and further development is needed before a new and highly interesting type of drug-delivery

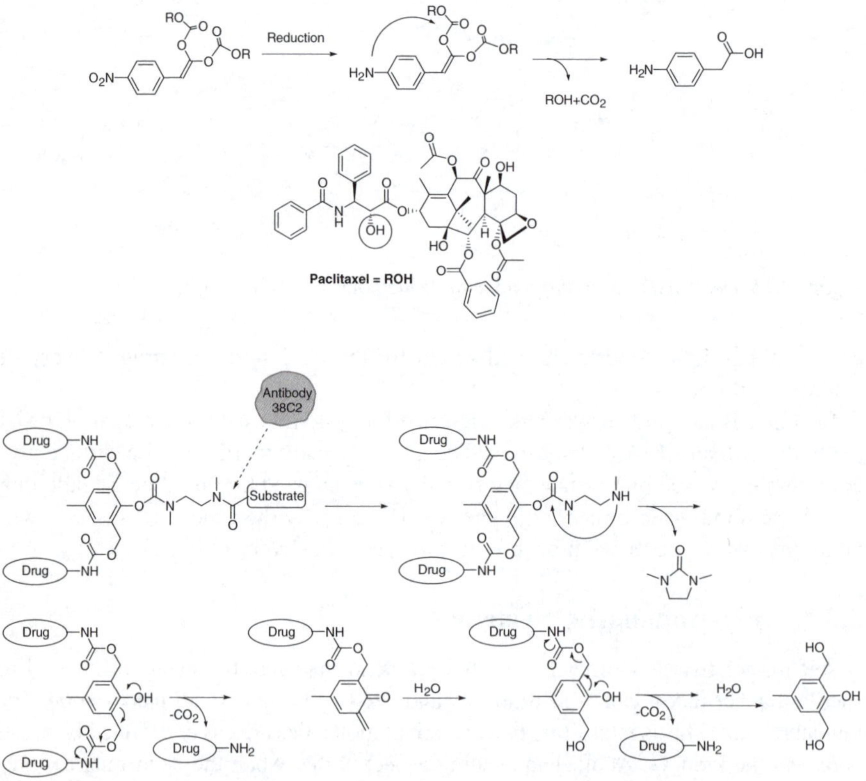

Figure 3.15 *Top: A model system capable of releasing Paclitaxel (R-OH) upon reduction of the nitro group at the focal point. The hydroxyl group marked with a circle is the attachment point to the linker. Bottom: Release of an amine-containing drug by an immolative cascade initialised by the hydrolase activity of the catalytic antibody 38C2*

system will emerge. The concept could provide an important tool for the treatment of cancer provided that systems triggered by enzymes present in the target cell can be developed.

3.4 Dendrimers as Gene Transfer Reagents

As transfection of eukaryotic cells is a methodology for effecting changes in the genetic material of cells, it has become a valuable tool in molecular biology for characterisation of eukaryotic cells, for studying mutations and regulation processes of genes or inducing overexpression of desired proteins.

Curing genetically based diseases by gene therapy is a goal of much interest. The ideal vector for transfection should apart from high efficiency be non-immunogenic, non-toxic, either biodegradable or excretable and preferably has a long blood circulation time.

Viruses are nature's own vehicles for this process, and much effort has been put into using this machinery provided by viruses for transfection of the cells. However, at least one case of death in a clinical trial has been reported[80,81] and other more technical problems are associated with the large-scale production and purification of viruses.[82–87]

The use of dendrimers for transfection was first reported by the groups of Szoka[88] and Baker.[89] They used commercial PAMAM-dendrimers, and found that these dendrimers form complexes with plasmid DNA (also coined polyplexes) capable of transfecting CV-1, HeLa, HepG2, Rat Hepatocyte, K562, EL-4 and Jukart cells. This was shown by the expression of the enzyme luciferase that originates from the firefly. The enzyme is easily detected via the light emission observed by an addition of luciferin to the transfected cells. The ability of PAMAM-dendrimers to form polyplexes with DNA was not totally unexpected, since polyamines such as polyethyleneimines are known agents for transfection. There is a clear correlation between the presence of non-titratable charges in the vector and transfection efficiency. For example, polylysine dendrimers, which have fixed non-titratable amino groups due to intramolecular protonisation of the amines by lysine acid residues, are a much less-efficient transfection agent than polyethyleneimine or a PAMAM-dendrimer. This leads to the proposal of the "proton sponge hypothesis" by Szoka[88] and Behr,[90] which postulates that the release of DNA from the DNA–vector complex in the endosome is due to the buffering properties of the vector leading to an accumulation of H^+ and subsequently Cl^- in the endosome resulting in an osmotic swelling/lysis of the endosome (Figure 3.16).

A part of this mechanism for transfection has been proven, the polyplex is taken into the cell by endocytosis, followed by an osmotic burst of the polyplex-containing endosome.

Experimental evidence for this mechanism was provided by the groups of George[91] and Verkman:[92] Eukaryotic cell membranes contain different types of lipids, cholesterol and proteins, and the properties of the membrane depend on the composition. Cholesterol is involved in endocytosis.[93,94] Cholesterol can be extracted from cell membranes by treatment with methyl-β-CD, which forms a guest–host complex with cholesterol. George and co-workers[91] investigated the effect of cholesterol removal

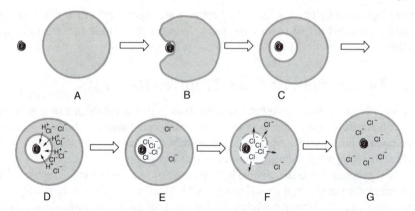

Figure 3.16 *Cellular uptake of DNA–dendrimer polyplexes by endocytosis. A: Cell. B: The endosome begins to fold around the polyplex. C: Endosome and the polyplex is internalised. D: Influx of H^+ to the endosome because of the overall basicity of the polyplex. E: Influx of Cl^-. F: The endosome breaks owing to osmotic swelling; and G: The DNA–dendrimer polyplex is released and the cell is transfected*

upon transfection of epithelial cells with DNA–dendrimer complexes (polyplexes). They found that the removal of cholesterol before transfection leads to poor expression, while there was no effect after transfection. Thus indicating that cholesterol is involved in the cellular uptake of dendrimers.[91]

Verkman and co-workers[92] devised a clever experiment to follow the uptake of the polyplex: they labelled a PAMAM-dendrimer, polyethyleneimine and polylysine with the different fluorescent dyes tetramethylrhodamine (TMR), fluoresceine isothiocyanate (FITC) and 10,10′-bis[3-carboxypropyl]-9,9′-biacridinium dinitrate (BAC). The fluorescence of TMR is not sensitive to the environment. The fluorescence of FITC is sensitive to pH whereas the fluorescence of BAC is sensitive to the chloride concentration. The vectors were labelled with a TMR label as control and either FITC or BAC. The DNA–vector complexes were compartmentalised inside the cell after an uptake. In the case of PAMAM or polyethyleneimine, there was a faster influx of H^+ to the endosome than in the case of polylysine. This influx of H^+ was followed by a similar influx of Cl^- finally leading to an increased osmotic pressure inside the endosome and lysis with a release of the content into the cytosol. How the DNA is transported from the cytosol and to the nucleus is not known. An interesting twist is the results reported by Yoo and Juliano[95] reported on the transfection of HeLa cells with fluorescent PAMAM-dendrimers. They found that the labelled dendrimer gave a better transfection than the unlabelled. This finding suggests that the use of dendrimers partially substituted with large aromatic groups might be more efficient. Further support comes from the work by Kono and co-workers,[96] who investigated transfection with PAMAM-dendrimers modified with leucine or phenylalanine at the periphery. They studied G4 dendrimers and found that the phenylalanine-substituted dendrimer gave a much better transfection than the leucine substituted. The "naked" Phe-dendrimer was more toxic than lipofectamine and SuperFect, but the DNA complex had less toxicity while the transfection activity was higher than that of lipofectamine and SuperFect. In their initial studies of transfection with PAMAM-dendrimers, Szoka and co-workers[97,98] discovered that the purity of the PAMAM-dendrimers was of crucial importance: but not in the normal manner, where higher purity gives better results. The best results were obtained with defect-containing PAMAM-dendrimers. In order to explain this observation, the so-called "umbrella-model" was put forward. A defect dendrimer missing a number of arms in the structure, makes it much more flexible than the perfect dendrimer. The complex between the DNA and the defect dendrimer undergoes a much larger conformational change upon protonation than the perfect dendrimer, and this makes the separation of the DNA from the dendrimer faster in the case of the defect dendrimer. This phenomenon can be compared to the unfolding of an umbrella.

A number of different dendrimers have been prepared and studied as vectors for transfection, and the following section is a short description of the different dendrimers that have been used:

PAMAM-dendrimers were the first type of dendrimers, which were found to be of useful for transfection. Following the pioneering works of Szoka *et al.*[88] and Baker *et al.*,[89] the company Quiagen developed a commercial transfection system based on PAMAM-dendrimers.[84]

Diederich and co-workers[99] described a series of amphiphilic dendrimers (Figure 3.17), and compared the transfection efficiency of the compounds with DOTAP (a lipid-based transfection agent from Roche), polyethyleneimine and SuperFect (a degraded PAMAM from Quiagen). They found that the optimal transfection efficiency was found with the two medium-sized dendrimers 2 and 3, which were both better than the known compounds tested. However in the presence of the serum, which has been known to affect transfection efficacy, compound 2 was much better than compound 3.

The use of phosphorous-based dendrimers for transfection of 3T3-cells with the luciferase gene has been investigated by Majoral, Caminade and co-workers.[100] They found some generation-dependence with the dendrimers, with the G4–G6 dendrimers being somewhat more efficient than G2 and G3. The G5 dendrimer had lower cytotoxicity and higher luciferase expression than polyethyleneimine, the transfection efficiency was not affected by the presence of 10% of serum. Zinselmeyer and co-workers[101] compared the transfection efficiency of all five generations of commercially available polypropyleneimine dendrimers with DOTAP in the A431 cell line, and found that all the compounds were more or less cytotoxic with the highest generation dendrimer as the most toxic compound. Interestingly, the G2 dendrimer was the least toxic compound. It was less toxic than DOTAP and had comparable transfection efficiency.

Inhibition of expression of the vascular endothelial growth factor (VEGF) involved in the development of blood vessels (angiogenesis, see also Chapter 6) was investigated in the complex formation between a dendrimer and an oligonucleotide called ODN-1 (sequence: GAGCCGGAGAGGGAGCGCGA) in human RPE, D407 cell line.[102] This oligonucleotide had earlier been shown to inhibit VEGF production. The dendrimer serves two purposes, it should enable transportation of ODN-1 into the cell and it should protect ODN-1 against nucleases.

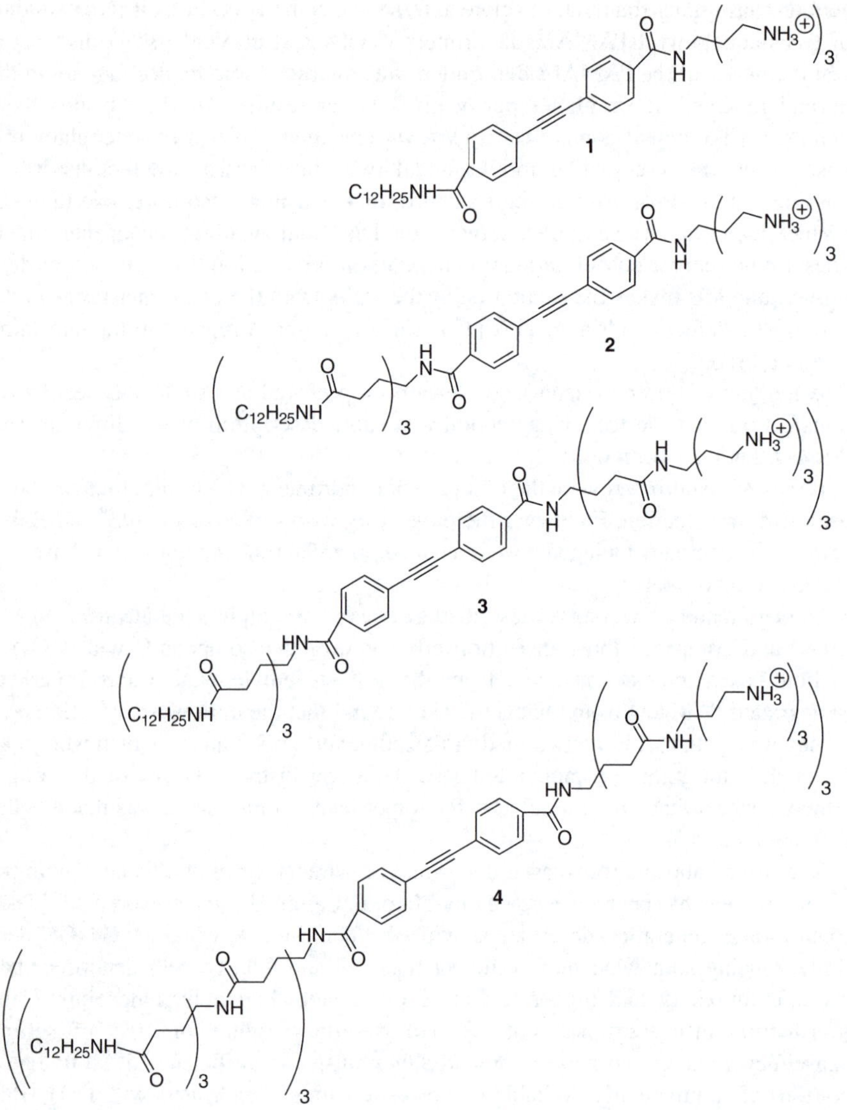

Figure 3.17 *Amphiphilic dendrimers developed for transfection by Diederich and co-workers.*[99] *The anion was CF₃COO⁻ for all the compounds shown*

VEGF is involved in age-related macular degeneration and diabetic retinopathy causing the slow loss of vision in diabetes patients, both ocular diseases being the main cause for blindness in the developed countries.[102,103]

The lysine-based generation 2 and 3 dendrimers used were modified on the C-terminal with lipid residues contaning oligopeptides made from 2-aminoalkanoic acids (2-aminododecanoic, 2-amino-tetradecanoic and 2-aminooctadecanoic acid) carrying a terminal glycinamide thereby varying the lipophilicity and the charge on the dendrimers. The interaction between ODN-1 and the dendrimers were studied by isothermal calorimetry (ITC), being a highly sensitive method for determination of stoichiometry as well as association constants for host–guest complexes. The overall charge of the dendrimers varied between +4 to +8, and it was found that the highest ODN-1 to dendrimer ratio (1:6) was obtained with dendrimers having a charge of 8^+ and either a C14 or C18-dipeptide as the lipophilic unit. These results correlated also with the transfection efficiency in cell cultures showing that the highly positively charged dendrimers are better for transfection. These dendrimers were tested in a rat model and it was found that the ODN-1–dendrimer complexes were active and kept their activity for up to two months.

3.5 Summary

Dendrimers are in general useful as a part of the molecular toolbox for drug delivery. Cellular uptake of dendrimers (and polymers) can take place by endocytosis in cells and thus brings the drugs imbedded in the dendrimer into the cell. Dendrimers show the EPR effect, which makes dendrimers attractive for targeting tumours, because tumour vasculature has increased permeability and limited lymphatic drainage leading to an accumulation of the dendrimers and their payload in the tumour.

Dendrimers are compartmentalised – it is a consequence of their spatial structure, and this gives different options in using them for drug delivery. Dendrimers are also suitable for tagging with molecules, which can bind to specific sites on cell surfaces making targeted delivery possible. This has been demonstrated with folic acid and with antibodies.

Suitably modified dendrimers can be used for transdermal drug delivery, which is an alternative to the oral administration, allowing direct delivery of the drug through the skin into the tissue.

Self-immolative dendrimers are a new type of dendrimers, which disassemble into small components on activation of a built-in trigger. These systems might be a valuable tool in future for chemotherapy in combination with a targeted drug delivery.

Transfection using dendrimers as vectors is an area, where dendrimers already have found uses, and have proven to be valuable alternatives to the classical transfection strategies. Compared to viruses, which are nature's own vehicles for this process, they are much more safe and the toxicity can be controlled and modulated via the structure of the dendrimer, making dendrimers promising candidates for future gene therapy.

To conclude this chapter, dendrimers of different shapes and structures can be used as drug-delivery vehicles, and many of the results reported so far point towards dendrimers as an upcoming important class of compounds in the delivery field.

References

1. H. Brem, K.A. Walter and R. Langer, *Eur. J. Pharm. Biopharm.*, 1993, **39**, 2.
2. H. Ringsdorf, *J. Polym. Sci. C*, 1975, **5**, 135.
3. G. Franzmann and H. Ringsdorf, *Macromol. Chem.-Macromol. Chem. Phys.*, 1976, **177**, 2547.

4. J. Kalal, J. Drobnik, J. Kopecek and J. Exner, *Brit. Polym. J.*, 1978, **10**, 111.
5. R. Duncan, *Anti-Cancer Drugs*, 1992, **3**, 175.
6. E.R. Gillies, E. Dy, J.M.J. Frechet and F.C. Szoka, *Mol. Pharm.*, 2005, **2**, 129.
7. R. Duncan and J. Kopecek, *Adv. Polym. Sci.*, 1984, **57**, 51.
8. F. Aulenta, W. Hayes and S. Rannard, *Eur. Polym. J.*, 2003, **39**, 1741.
9. K. Uhrich, *Trends Polym. Sci.*, 1997, **5**, 388.
10. A. Nori and J. Kopecek, *Adv. Drug Delivery Rev.*, 2005, **57**, 609.
11. S.E. Stiriba, H. Frey and R. Haag, *Angew. Chem. Int. Ed. Engl.*, 2002, **41**, 1329.
12. R. Duncan and F. Spreafico, *Clin. Pharm.*, 1994, **27**, 290.
13. D.G. Kanjickal and S.T. Lopina, *Crit. Rev. Ther. Drug Carrier Sys.*, 2004, **21**, 345.
14. R.K.Y. Zeecheng and C.C. Cheng, *Methods Findings Exp. Clin. Pharm.*, 1989, **11**, 439.
15. A.K. Patri, I.J. Majoros and J.R. Baker, *Curr. Opinion Chem. Biol.*, 2002, **6**, 466.
16. N. Malik, R. Wiwattanapatapee, R. Klopsch, K. Lorenz, H. Frey, J.W. Weener, E.W. Meijer, W. Paulus and R. Duncan, *J. Cont. Rel.*, 2000, **65**, 133.
17. P.L. Rinaldi, *Analyst*, 2004, **129**, 687.
18. K. Schmidt-Rohr and H.W. Spiess. *Ann. Rep. NMR Spect.*, 2002, **48**, 1.
19. G.R. Newkome, C.N. Moorefield, G.R. Baker, A.L. Johnson and R.K. Behera, *Angew. Chem. Int. Ed. Engl.*, 1991, **30**, 1176.
20. G.R. Newkome, C.N. Moorefield, G.R. Baker, M.J. Saunders and S.H. Grossman, *Angew. Chem. Int. Ed. Engl.*, 1991, **30**, 1178.
21. G.R. Newkome, X.F. Lin, C. Yaxiong and G.H. Escamilla, *J. Org. Chem.*, 1993, **58**, 3123.
22. C.N. Moorefield and G.R. Newkome, *Compt. Rend. Chim.*, 2003, **6**, 715.
23. S. Stevelmans, J.C.M. van Hest, J.F.G.A. Jansen, D.A.F.J. van Boxtel, E.M.M.D. van den Berg and E.W. Meijer, *J. Am. Chem. Soc.*, 1996, **118**, 7398.
24. M.W.P.L. Baars, P.E. Froehling and E.W. Meijer, Chem. Commun., 1997, 1959.
25. M.W.P.L. Baars, R. Kleppinger, M.H.J. Koch, S.L. Yeu and E.W. Meijer, *Angew. Chem. Int. Ed. Engl.*, 2000, **39**, 1285.
26. C.J. Hawker, K.L. Wooley and J.M.J. Frechet, *J. Chem. Soc. Perkin Trans. 1*, 1993, **12**, 1287.
27. M.J. Liu, K. Kono and J.M.J. Frechet, *J. Cont. Rel.*, 2000, **65**, 121.
28. D.M. Watkins, Y.S. Sweet, J.W. Klimash, N.J. Turro and D.A. Tomalia, *Langmuir*, 1997, **13**, 3136.
29. D. Bhadra, S. Bhadra, S. Jain and N.K. Jain, *Int. J. Pharm.*, 2003, **257**, 111.
30. L.J. Twyman, A.E. Beezer, R. Esfand, M.J. Hardy and J.C. Mitchell, *Tetrahedron Lett.*, 1999, **40**, 1743.
31. A.E. Beezer, A.S.H. King, I.K. Martin, J.C. Mitchel, L.J. Twyman and C.F. Wain, *Tetrahedron*, 2003, **59**, 3873.
32. W. Zhang, J. Jiang, C.H. Qin, L.M. Perez, A.R. Parrish, S.H. Safe and E.E. Simanek, *Supramol. Chem.*, 2003, **15**, 607.
33. H.T. Chen, M.F. Neerman, A.R. Parrish and E.E. Simanek, *J. Am. Chem. Soc.*, 2004, **126**, 10044.
34. M.F. Neerman, W. Zhang, A.R. Parrish and E.E. Simanek, *Int. J. Pharm.*, 2004, **281**, 129.

35. A.S. Chauhan, S. Sridevi, K.B. Chalasani, A.K. Jain, S.K. Jain, N.K. Jain and P.V. Diwan, *J. Cont. Rel.*, 2003, **90**, 335.
36. A.S. Chauhan, N.K. Jain, P.V. Diwan and A.J. Khopade, *J. Drug Target.*, 2004, **12**, 575.
37. T.F. Vandamme and L. Brobeck, *J. Cont. Rel.*, 2005, **102**, 23.
38. R.F. Barth, D.M. Adams, A.H. Soloway, F. Alam and M.V. Darby, *Bioconj. Chem.*, 1994, **5**, 58.
39. N. Malik, E.G. Evagorou and R. Duncan, *Anti-Cancer Drugs*, 1999, **10**, 767.
40. J.F.G.A. Jansen, E.M.M. de Brabander van den Berg and E.W. Meijer, *Science*, 1994, **266**, 1226.
41. J.F.G.A. Jansen, E.W. Meijer and E.M.M. de Brabander van den Berg, *Macromol. Symp.*, 1996, **102**, 27.
42. J.F.G.A. Jansen, E.M.M. de Brabander van den Berg and E.W. Meijer, *Abstr. Pap. Am. Chem. Soc.*, 1995, **210**, 64-MSE.
43. A.W. Bosman, J.F.G.A. Jansen, R.A.J. Janssen and E.W. Meijer, *Abstr. Pap. Am. Chem. Soc.*, 1995, **210**, 181-MSE.
44. J.F.G.A. Jansen, E.M.M. de Brabander van den Berg and E.W. Meijer, *Rec. Trav. Chim. Pays-Bas*, 1995, **114**, 225.
45. J.F.G.A. Jansen, R.A.J. Janssen, E.M.M. de Brabander van den Berg and E.W. Meijer, *Adv. Mat.*, 1995, **7**, 561.
46. J.F.G.A. Jansen, E.W. Meijer and E.M.M. de Brabander van den Berg, *J. Am. Chem. Soc.*, 1995, **117**, 4417.
47. L. Balogh, D.R. Swanson, D.A. Tomalia, G.L. Hagnauer and A.T. McManus, *Nano Lett.*, 2001, **1**, 18.
48. P. Wallimann, P. Seiler and F. Diederich, *Helv. Chim. Acta*, 1996, **79**, 779.
49. D.K. Smith, A. Zingg and F. Diederich, *Helv. Chim. Acta*, 1999, **82**, 1225.
50. D.K. Smith and F. Diederich, *Chem. Commun.*, 1998, 2501.
51. G.R. Newkome, B.D. Woosley, E. He, C.N. Moorefield, R. Guether, G.R. Baker, G.H. Escamilla, J. Merrill and H. Luftmann, *Chem. Commun.*, 1996, 2737.
52. M. Puri, S.S. Marwaha, R.M. Kothari and J.F. Kennedy, *Crit. Rev. Biotech.*, 1996, **16**, 145.
53. A.P. Sayani and Y.W. Chien, *Crit. Rev. Ther. Drug Carrier Sys.*, 1996, **13**, 85.
54. E. Albers and B.W. Muller, *Crit. Rev. Ther. Drug Carrier Sys.*, 1995, **12**, 311.
55. B. Gonzáles, C.M. Casado, B. Alonso, I. Cuadrado, M. Morán, Y. Wang and A.E. Kaifer, *Chem. Commun.*, 1998, 2569.
56. M.W.P.L. Baars, E.W. Meijer, J. Huskens, D.N. Reinhoudt and J.J. Michels, *J. Chem. Soc. Perkin Trans. 2*, 2000, 1914.
57. P. Kolhe, E. Misra, R.M. Kannan, S. Kannan and M. Lieh-Lai, *Int. J. Pharm.* 2003, **259**, 143.
58. H. Stephan, H. Spies, B. Johannsen, L. Klein and F. Vogtle, *Chem. Commun.*, 1999, 1875.
59. J.C. Sisson, *Thyroid*, 1997, **7**, 295.
60. M.W.P.L. Baars and E.W. Meijer, *Top. Curr. Chem.* 2000, **210**, 131.
61. U. Boas, A.J. Karlsson, B.F.M. de Waal and E.W. Meijer, *J. Org. Chem.*, 2001, **66**, 2136.

62. U. Boas, S.H.M. Sontjens, K.J. Jensen, J.B. Christensen and E.W. Meijer, *Chembiochem*, 2002, **3**, 433.
63. M. Pittelkow, J.B. Christensen and E.W. Meijer, *J. Polym. Sci. Part A*, 2004, **42**, 3792.
64. A. Dirksen, U. Hahn, F. Schwanke, M. Nieger, J.N.H. Reek, F. Vogtle and L. De Cola, *Chem. Eur. J.*, 2004, **10**, 2036.
65. R.X. Zhuo, B. Du and Z.R. Lu, *J. Cont. Rel.*, 1999, **57**, 249.
66. R. Wiwattanapatapee, L. Lomlim and K. Saramunee, *J. Cont. Rel.*, 2003, **88**, 1.
67. J. Khandare, P. Kolhe, O. Pillai, S. Kannan, M. Lieh-Lai and R.M. Kannan, *Bioconj. Chem.*, 2005, **16**, 330.
68. R. Jevprasesphant, J. Penny, R. Jalal, D. Attwood, N.B. McKeown and A. D'Emanuele, *Int. J. Pharm.*, 2003, **252**, 263.
69. R. Jevprasesphant, J. Penny, D. Attwood and A. D'Emanuele, *J. Cont. Rel.*, 2004, **97**, 259.
70. R. Jevprasesphant, J. Penny, D. Attwood, N.B. McKeown and A. D'Emanuele, *Pharm. Res.*, 2003, **20**, 1543.
71. A. D'Emanuele, R. Jevprasesphant, J. Penny and D. Attwood, *J. Cont. Rel.*, 2004, **95**, 447.
72. K. Kono, M.J. Liu and J.M.J. Frechet, *Bioconj. Chem.*, 1999, **10**, 1115.
73. A. Quintana, E. Raczka, L. Piehler, I. Lee, A. Myc, I. Majoros, A.K. Patri, T. Thomas, J. Mule and J.R. Baker, *Pharm. Res.*, 2002, **19**, 1310.
74. A.K. Patri, A. Myc, J. Beals, T.P. Thomas, N.H. Bander and J.R. Baker, *Bioconj. Chem.*, 2004, **15**, 1174.
75. R.J. Amir, N. Pessah, M. Shamis and D. Shabat, *Angew. Chem. Int. Ed. Engl.*, 2003, **42**, 4494.
76. F.M.H. de Groot, C. Albrecht, R. Koekkoek, P.H. Beusker and H.W. Scheeren, *Angew. Chem. Int. Ed. Engl.*, 2003, **42**, 4490.
77. M.L. Szalai, R.M. Kevwitch and D.V. McGrath, *J. Chem. Soc. Perkin Trans. 1*, 2003, **125**, 15668.
78. R.J. Amir and D. Shabat, *Chem. Commun.*, 2004, 1614.
79. M. Shamis, H.N. Lode and D. Shabat, *J. Am. Chem. Soc.*, 2004, **126**, 1726.
80. Anon., *Science*, 2001, **294**, 1638.
81. N. Somia and I.M Verma, *Nat. Rev. Gen.*, 2000, **1**, 91.
82. G.D. Schmidt-Wolf and I.G.H. Schmidt-Wolf, *Trends Mol. Med.*, 2003, **9**, 72.
83. D. Simberg, S. Weisman, Y. Talmon and Y. Barenholz, *Crit. Rev. Ther. Drug Carrier Sys.*, 2004, **21**, 257.
84. J. Dennig, *Top. Curr. Chem.*, 2003, **228**, 227.
85. A.D. Miller, *Curr. Med. Chem.*, 2003, **10**, 1195.
86. C.M. Middaugh and C.R. Wiethoff, *J. Pharm. Sci.*, 2003, **92**, 203.
87. T. Niidome and L. Huang, *Gene Ther.*, 2002, **24**, 1647.
88. J. Haensler and F.C. Szoka, *Bioconj. Chem.*, 1993, **4**, 372.
89. J.F. KukowskaLatallo, A.U. Bielinska, J. Johnson, R. Spindler, D.A. Tomalia and J.R. Baker, *Proc. Nat. Acad. Sci. USA*, 1996, **93**, 4897.
90. J.P. Behr, *Bioconj. Chem.*, 1994, **5**, 382.
91. M. Manunta, P.H. Tan, P. Sagoo, K. Kashefi and A.J.T. George, *Nucl. Acids Res.*, 2004, **32**, 2730.

92. N.D. Sonawane, F.C. Szoka and A.S. Verkman, *J. Biol. Chem.*, 2003, **278**, 44826.
93. H.J. Heiniger, A.A. Kandutsch and H.W. Chen, *Nature*, 1976, **263**, 515.
94. R. Wattiaux, N. Laurent, S.W.-D. Coninck and M. Jadot, *Adv. Drug Delivery Rev.*, 2000, **41**, 201.
95. H. Yoo and R.L. Juliano, *Nucl. Acids Res.*, 2000, **28**, 4225.
96. K. Kono, H. Akiyama, T. Takahashi, T. Takagishi and A. Harada, *Bioconj. Chem.*, 2005, **16**, 208.
97. M.X. Tang, C.T. Redemann and F.C. Szoka, *Bioconj. Chem.*, 1996, **7**, 703.
98. M.X. Tang and F.C. Szoka, *Gene Ther.*, 1997, **4**, 823.
99. D. Joester, M. Losson, R. Pugin, H. Heinzelmann, E. Walter, H.P. Merkle and F. Diederich, *Angew. Chem. Int. Ed. Engl.*, 2003, **42**, 1486.
100. C. Loup, M.A. Zanta, A.M. Caminade, J.P. Majoral and B. Meunier, *Chem. Eur. J.*, 1999, **5**, 3644.
101. B.H. Zinselmeyer, S.P. Mackay, A.G. Schatzlein and I.F. Uchegbu, *Pharm. Res.*, 2002, **19**, 960.
102. R.J. Marano, N. Wimmer, P.S. Kearns, B.G. Kearns, B.G. Thomas, I. Toth, M. Brankow and P.E. Racoczy, *Exp. Eye Res.*, 2004, **79**, 525.
103. N. Wimmer, R.J. Marano, P.S. Kearns, E.P. Rakoczy and I. Toth, *Bioorg. Med. Chem. Lett.*, 2002, **12**, 2635.

CHAPTER 4

Dendrimer Drugs

4.1 Introduction

In addition to the ability to carry drugs as described in the previous chapter, dendrimers can also be synthesised to possess intrinsic biological properties that can be useful for treatment and other medical interventions. Dendrimers offer a range of advantages with respect to mediating biological effects in the living organism: (1) a high degree of molecular definition; (2) versatility of design; and (3) multivalent presentation of a certain motif. Therefore dendrimers have a surprisingly wide scope of applications, the possibilities of which have only just been begun to be investigated. Dendrimer drugs are cost efficient, generally being produced by standard chemical reactions and they allow the use of simple compounds that may work biologically because they are presented multivalently by the dendrimer, and even, by targeting the dendrimer to the tissue/cell in question can be presented to the target at a very high local concentration. With dendritic drugs, it is possible to exploit the multivalency of the dendrimer to achieve an enhanced effect of the drug. The multivalent presentation may allow the use of simple compounds that will work biologically because they are presented in a multivalent fashion, exploiting the synergy effect (the "dendritic effect", see Chapter 2) to achieve adequate affinities and biological potency. This principle has especially been used with carbohydrates where simple saccharides presented on a dendrimer are able to mimic far more complicated carbohydrate monomers (*vide infra*) or may enhance efficiency of binding to a carbohydrate receptor significantly.[1,2]

The so-called "disassembling" dendrimers, also called "cascade release" or "self-immolative" dendrimers,[3-5] are examples of substances that are intermediate between drugs being carried by dendrimers (Chapter 3) and the dendrimer drugs described in the present chapter. Such dendrimers comprise a "trigger" that mediates the release of bioactive unit(s) from the dendrimer and causes the whole dendrimer to fragment.[6] This renders the local concentration of the released drug very high and therefore efficient, and the dendrimer fragments are easily cleared by the body thus avoiding the toxicity of the "naked" dendrimer often seen when using dendrimers as drug carriers.

As for drug delivery, the *in vivo* use of dendrimer drugs depends on its pharmacokinetics, the stability of the molecule, and its non-immunogenicity (if not intended as a vaccine). Other central challenges for the use and design of dendrimers as drugs in the living organism include the following:

- to establish the right balance between the desired biological effect and the general toxic effect (typically cytotoxicity and haemolytic effects) of the dendrimer;
- to define the optimal size in order for the dendrimer drug to possess the right binding affinity and spatial dimensions in interactions with receptors and in order for the dendrimer drug to be able to penetrate the physiological barriers and cell membranes that are relevant to its desired biological effects;
- to ensure the optimal multivalency in order to achieve the affinity needed for the drug to interact as desired with its target;
- to render the dendrimer drug soluble in physiological conditions of pH and ionic strength; and
- to stabilise the dendrimer drug adequately so that it will survive for sufficient time in the relevant body compartments.

As a generic platform technology, dendrimers allow for all of these points to be addressed through their flexibility of design.

As discussed in Chapter 2, delivery of dendrimer drugs by the circulatory system (intravascular delivery) is complicated by non-specific binding to both soluble and particulate components of the blood. Also, targetting the dendrimer to the site of action – which may be extravascular – is a major challenge. Moreover, dendrimers may rapidly be removed from the blood stream, often within an hour.[7] It is known from work with dendrimer-based contrast agents[8] (see Section 5.1, Chapter 5) that PPI dendrimers are more rapidly excreted from the body (through the kidneys) than PAMAM dendrimers of the same generation and that small cationic dendrimers are more rapidly excreted by the renal route than larger dendrimers. Cationic dendrimers (full-generation PAMAM) stayed shorter in circulation than comparable anionic dendrimers (half-generation PAMAM) and for anionic dendrimers lower generations were removed from the circulation more slowly than higher generations.[7]

All of this is of importance for a dendrimer drug to be able to interact with its target. Dendrimers also have superior membrane penetrating properties and can reach targets inside cells.[9] The ability of dendrimers to interact with membranes can be utilised in the design of dendrimer-based biocides, but can also be a complicating factor leading to excessive cytotoxicity as discussed elsewhere in this volume (see Chapter 2). By the right combination of generation number, solubility and surface charges these kinds of side effects can be counteracted. Cationic dendrimers are generally more cytotoxic and haemolytic than anionic dendrimers, and small generation dendrimers are generally less toxic than higher generations.[7]

4.2 Antiviral Dendrimers

Viruses work by binding to the host cell surface (adhesion) followed by internalisation by the cell (penetration/entry), whereafter the virus uses the protein-synthesising and DNA-replicating machinery of the host cell to multiply itself (viral genome and coat). Viruses do not respond to antibiotics, immunity may be hard to come by, especially when dealing with viruses that attack host cells belonging to the immune system (HIV being an important example) and with quickly mutating viruses, and viral infections often pave the way for other (bacterial) infections by weakening the host leading to a more severe clinical situation for the infected host.

Dendrimers, which are only a little smaller than most viruses (Table 1.1, Chapter 1) can be designed to interfere with the virus-to-host cell binding process thereby inhibiting infection at the stage of viral entry as an efficient, early interference with the infection. The dendrimer can be designed to either bind specific receptors on the host cell surface or to bind to viral surface components. It can also be built to bind generally to the virus surface through electrostatic forces (typically anionic dendrimers). Blockade of receptor binding can be difficult to achieve as several receptors with varying degrees of involvement in adhesion and penetration events may be involved.

To study the effect of such drugs, model systems are often used, including binding assays using purified target cells or target cell components and studying the binding of the virus to this in the presence of potentially inhibiting drugs. This gives information about the ability of the drug to inhibit the binding of the virus to the cell, often expressed as IC50, *i.e.* the concentration of the drug needed to give a 50% reduction of the maximum binding. Inhibition of infectivity can be studied in cell culture as can interference with virus in already infected cells which will be of importance with regard to the ability of the dendrimer drug to be able to treat already established infections (which is by far the most common real life situation). Finally, drugs are tested in *in vivo* models (Figure 4.1).

Antiviral dendrimers working as artificial mimics of the target cell surface are generally designed with anionic surface groups like sulfonate residues or sialic acid residues mimicking the acidic carbohydrates present at the mammalian cell surface. The polyanionic dendritic drug then competes with the cellular surface for binding of virus, leading to a lower cell-virus infection rate (Figure 4.2). However, as we shall see in the following chapter, antiviral dendrimers may also be designed having cationic or neutral surfaces, depending on their mode of action.

Concerning the important first events leading to infection by virus, several types of drugs have been shown to inhibit adhesion. With herpes simplex virus (HSV) both polycationic compounds (polyarginine and polylysine)[10] and polyanionic compounds,[11] including polyanionic dendrimers[12,13] could inhibit adsorption of the virus to cell surfaces; this can be explained by an antagonistic effect either competing with the virus for a cell-associated anionic receptor structure or by competing with the cell for a cationic virus component. However, the drug quality of polycationic dendrimers is hampered by their higher cytotoxicity in comparison to polyanionic dendrimers.[7]

In a study by Reuter and co-workers,[14] sialic acid-terminated dendrimers were used to inhibit attachment of influenza virus to sialic acid containing glycoproteins on the surface of the target cells, which are the natural receptors for the influenza virus haemagglutinin spike proteins on the influenza virion. In the opposite approach, a target cell lectin was blocked in its binding to virus envelope glycoproteins using mannosylated dendrimers to block binding of ebola envelope glycoprotein to the dendritic cell C-type lectin.[15] The late effect observed in these systems is

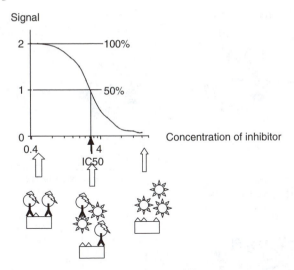

Figure 4.1 *The principle of determining IC50 as an expression of inhibitory potential of a dendrimer drug. IC50 is the concentration of the dendrimer inhibitor resulting in a decrease of the signal due to bound signal generation entity (for example, a labelled virus) to 50% of the max signal. The rest of the signal is washed away from the solid phase carrying the ligand*

probably owing to uptake of the dendrimer into virus-infected cells where it inhibits virus replication by an unknown mechanism.[12,16] There is no evidence of preferential uptake in virus infected cells as opposed to non-infected cells.

A G3 lysine-derived dendrimer with sulfonated naphthalene groups coupled to the surface amino groups through an amide bond is a polyanionic dendrimer drug studied by several groups for interference with virus binding and infection.[12,13,17] These dendrimers inhibit adhesion of HSV *in vitro* and were also tested unformulated and formulated in various gels *in vivo* and provided protection against genital herpes in mice and guinea pigs for up to 1 h after intravaginal application. Thus, dendrimers like these offer the possibility of using molecularly defined microbicides in contrast to the commonly used compounds, which are typically undefined mixtures with some side effects; although dendrimers may not be superior to drugs in use in this field, their high definition gives certain regulatory/approval benefits. Formulations can still be optimised to optimise adhesion (*i.e.* time of action *in situ*), pH, *etc.*, but the results hold promise for the treatment of other sexually transmitted viral and bacterial infections like HIV and chlamydia. Early and later stages of virus replication were both shown to be inhibited in the cell culture model of HSV replication[12] using 32-surface group lysine-based dendrimer decorated with 3,6-disulfonyl naphthalene groups;[17] this was evidenced by inhibition of the cytopathic effect of the virus (inhibition of adsorption and entry) in addition to inhibition of DNA synthesis in already infected cells (late stage inhibition) as shown in a separate experiment. This worked with both HSV-1 and -2.

These anionic dendrimers were further studied by Jiang and co-workers[18] who investigated chimeric simian/HIV transmission in macaques and its inhibition by a

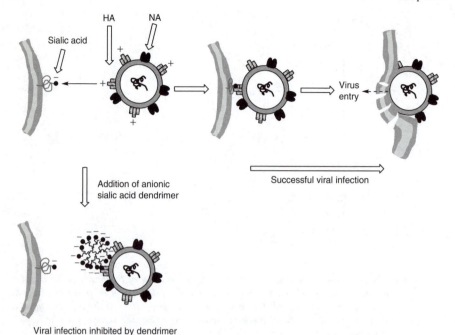

Figure 4.2 *Schematic showing the interaction of the influenza virus surface receptor (the haemagglutinin, HA) for sialic acid (N-acetyl neuraminic acid, NaNa⁻), mediating entry of the virus through binding to cell-surface glycoconjugates carrying oligosaccharides terminating in sialic acid (which is a frequent feature of eukaryotic cell surface oligosaccharides). When the virus neuraminidase (NA) cleaves off the sialic acid, the cell-surface oligosaccharide does not bind the virus HA anymore which allows virus "progeny" to leave the infected cell*

topical microbicide containing the same type of polyanionic dendrimer as above. Animals treated with the dendrimer showed a dose-dependent resistance against vaginal transmission of the virus with no adverse effects and 5% (50 mg mL^{-1} dendrimer) microbicide gels giving 100% protection. Only very high doses (above the 5% range) of the dendrimer exhibited cytotoxicity.

PAMAM dendrimers covalently modified with sulfonated naphthalene groups as above also showed antiviral activity against HIV. Also, in this case the dendrimer drug inhibited early stage virus/cell adsorption and at later stages of viral replication interfered with the reverse transcriptase and/or integrase enzymes.[16,19]

Dendrimers with non-charged surfaces have also been described as antiviral drugs; a dendrimer with an amide surface was shown to work as an inhibitor for the respiratory syncytial virus (RSV) (Figure 4.3). The exact mechanism of action has not been elucidated in detail, but may rely on hydrogen bonding interactions between the viral fusion protein and the dendrimer surface groups causing inhibition of virus binding and, especially, the fusion step following adsorption.[20] Even small alterations at the aromatic residues of the dendrimer decrease the antiviral activity and viral selectivity, suggesting that other binding modes than electrostatic binding

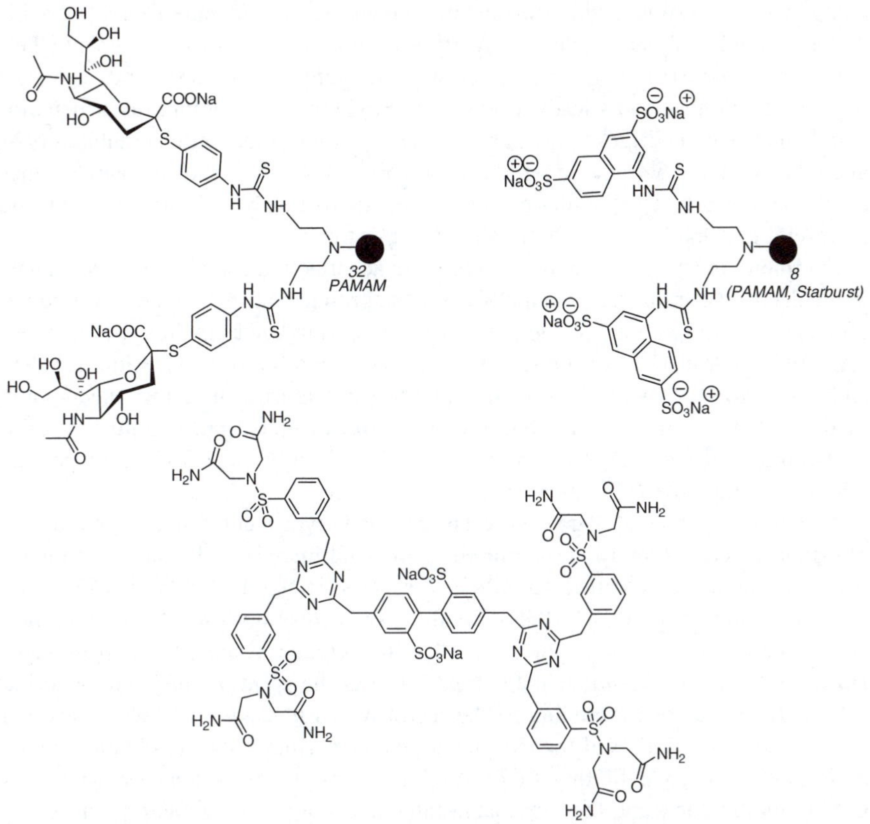

Figure 4.3 *Examples of molecular structures of antiviral dendrimers. Top left: antiviral dendrimer, carrying sialic acid on its surface, reactive against Influenza Pneumonitis.[23] Top right: antiviral dendrimer acting against HSV.[17] Bottom: antiviral dendrimer reactive against respiratory syncytial virus[21]*

e.g. π–π stacking could play a role as well.[21] Some of these compounds are very efficient having IC50s below 50 nM and inhibiting all types of RSV viruses and also work *in vivo* (rats and primates).[20]

In another approach, a G5 PPI dendrimer was coupled with polysulfated galactose to yield an inhibitor of HIV-1 as tested by an *in vitro* cell assay.[22] This was expected to work as an antagonist of HIV-1 binding to the target cells, as 3' sulfated galactosyl ceramide is known to function as a coreceptor for HIV-1. It was found that, in contrast to monosulfated galactose saccharides, the sulfated galactose dendrimer with 64 surface groups inhibited infection efficiently at the same level as sulfated dextran. Glycodendrimers with a range of different polyamide scaffolds, all terminated with sialic acid were tested for their ability to inhibit adhesion to red blood cells (haemagglutination assay) of different types of influenza viruses and a sendai virus.[14] Interestingly, cytotoxicity was inversely correlated with the degree of sialic acid substitution. Highly sialic acid-substituted dendrimers had a negligible cytotoxicity compared to dendrimers with low substitution with sialic acid and linear sialic acid-polyacrylamide polymers, which were efficient inhibitors of virus attachment albeit highly cytotoxic. This is in accordance with the general observation that a higher density of anionic groups leads to lower cytotoxicity.[14] They found linear-dendron copolymers and dendrigraft polymers to be 50,000 times more efficient inhibitors of agglutination than monomeric sialic acid and 500 times more efficient than the other dendrimers tested. This suggested that larger, more flexible dendrimers could be more efficient inhibitors of the relatively big influenza virus (120 nm diameter). Furthermore, the degree of sialylation or more accurately the number of free amines in the dendrimer influenced the inhibition – the presence of free amines lowered the efficiency of inhibition considerably. Partially sialylated, hydroxyl-terminated PAMAM dendrimers were found not to be able to inhibit virus adhesion. Landers and co-workers[23] brought the principle into use in a murine influenza pneumonitis model and could show the inhibition of influenza-mediated pneumonitis by sialic acid-conjugated G4 PAMAM dendrimers, although only the H3N2 sub-type was inhibited and not the H2N2 sub-type.

An interesting general target cell receptor is the C-type lectin found on the surface of dendritic cells (not to be confused with dendrimers!) and binding mannose residues. This lectin is called dendritic cell-specific intercellular adhesion molecule 3-grabbing non-integrin (DC-SIGN). Dendritic cells are important antigen-presenting cells functioning as early initiators of host immune responses to pathogens. However, some viruses utilise the C-type lectin as the point of entry into dendritic cells initiating entry by binding to the lectin by virus surface carbohydrates and therefore the application of mannosylated glycodendrimers was an obvious choice as an adhesion/entry inhibiting drug.[24] A G3 Boltorn-type (polyester) dendrimer was used to inhibit entry and the subsequent infection of a pseudo ebola retrovirus carrying the ebola envelope glycoprotein into dendritic cells and was shown to be able to do so at nanomolar concentrations.[15] Ebola virus is one of the most lethal pathogens known. It was first described in 1976 and natural outbreaks have hitherto been limited to isolated areas in Africa. It causes haemorrhagic fever after infection, is transmitted by direct contact with infected body fluids and has a mortality rate of 50–90%. There is no specific cure and no vaccine available, and there is a risk that ebola virus could be used for bioterrorism. A drug for treatment of the acute infection is therefore of great interest.[25]

This is also of some general interest as other important viruses, including HIV and cytomegalovirus (a herpes virus silently infecting a big part of the human world population giving rise to clinical disease in immunocompromised individuals only), utilise the DC-SIGN lectin to mediate entry into dendritic cells. Thus, drugs that interfere with the binding of DC-SIGN to its ligands may have many applications.

As mentioned above, in some cases dendrimers inhibit initial attachment of a certain virus to a cell surface and also have the ability to interfere with later stages of the infection process, *e.g.* HIV reverse transcriptase inhibition and HSV late stage replication.[12,16] Genital herpes, where the use of topical drugs is indicated have been investigated *in vitro* and *in vivo* (mice and guinea pigs) and a number of highly anionic dendrimers carrying sulfonate groups at the surface were shown to protect against herpes simplex transmission when applied intravaginaly at high concentrations.[13]

In conclusion, dendrimers have been designed to interfere with binding and *in vitro* infectivity of several viruses and in several cases were found to be able to inhibit infection efficiently either by inhibiting the binding of the virus to its target cell directly by binding to either the virus or the cell surface or by binding specific receptors on the virus surface. In the latter case, virus-type specific variations with respect to inhibition potential were found *e.g.* for influenza virus variants. Only a few reports have been published on *in vivo* antiviral applications of dendrimers and all of these have been within the area of topical viruses including herpes and using the dendrimer in a high concentration as part of a protective adhesive gel for protection against genitally transmitted virus infections.

4.3 Antibacterial Dendrimers

Traditional antibiotics have served the world exceptionally well in treating and controlling bacterial infections, but with an increasing incidence of antibiotics resistance among human and animal bacteria, new bacteriocidic drugs are highly needed.[26] Among candidates for new types of bacteriocides are a group of natural antimicrobial peptides which are found in eukaryotic organisms where they constitute a part of the innate or non-adaptive immune system, *i.e.* the defence system that recognises conserved microbial structures (danger signals) and counteracts microbial pathogens as a first line of defence.[27] A large group of these antimicrobial peptides has been isolated and characterised, including cecropins, defensins, cathelicidins and protgerins (see *e.g.* review by Reddy and co-workers[28]). These peptides share certain common features: (1) they are relatively small, (2) they are typically cationic, (3) they have hydrophobic domains distributed to make them amphipathic, and (4) their mode of action is to interact with and perturb the outer membranes of both gram negative and gram positive bacteria. Here, the positive charge is thought to play an important role in mediating the binding of the peptide to the highly negatively charged bacterial cell surface (which is very rich in acidic phospholipids) and the amphipathy is thought to drive the membrane perturbation, resulting in pore formation leading to bacterial cell lysis. One possible mechanism[28] is the barrel stave mechanism in which an aqueous pore is formed by the formation of a bundle of amphipathic peptides (typically α-helical), hydrophobic faces outwards and interacting with the membrane and the hydrophilic faces lining the central pore, leading to a leaking cell membrane, resulting in lysis. Another model is the carpet model in which the monomer peptide binds to the negatively charged phospholipids of the membrane and then in a cooperative process engulfs a piece of the membrane in the shape of a mixed membrane/peptide micelle. In contrast to bacterial cells, mammalian cells do not show anionic phospholipids on the outer surface of the cell membrane, instead they are positioned on the inner (cytoplasmic) face of the membrane. This may explain the higher reactivity of cationic peptides and dendrimers with microbial membranes. The composition of the membrane, *i.e.* the type of phospholipid buiding up the membrane also plays a role (Figure 4.4).[28]

Antibacterial dendrimers directed towards lysing bacteria thus have often been designed to carry cationic surface groups, one important point being to balance the size of the dendrimer and the number of cationic groups to achieve high activity

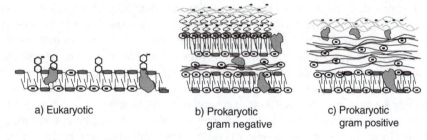

a) Eukaryotic   b) Prokaryotic   c) Prokaryotic
                                  gram negative   gram positive

Figure 4.4 *Comparison between the negative charge distribution on the typical eukaryotic membrane (a) and two common types of bacterial membranes (b, c). For all types of membranes negative charges dominate, but prokaryotic membranes carry more negative charges, especially on the outside of the membrane than does eukaryotic membranes. Encircled minus signs indicate that the negative charge is always found in the locations depicted, while a minus sign without a circle indicates that a negative charge __may__ be found in these locations too*

against prokaryotic membranes while keeping eukaryotic cytotoxicity at a minimum. As mentioned above, the outside of eukaryotic cell membranes is generally less negatively charged than prokaryotic membranes and contain other types of phospholipids that stabilise the membrane in a different way than found in prokaryotic membranes. Some of the antimicrobial peptides mentioned above lyse liposomes (artificial, membrane-surrounded vesicles) incorporating typical prokaryotic phospholipids like phosphatidylglycerol, but do not lyse such liposomes when incorporating phosphatidylserine, which is a typical eukaryotic phospholipid.[29]

While antibacterial dendrimers as a rule are cationic, model studies with anionic model lipid bilayers in water have shown that both full-generation (amine surface groups) and half-generation (carboxylate surface groups) PAMAM dendrimers are membrane disruptive, causing the formation of holes the size of the dendrimer itself in the membrane, while acetamide-terminated PAMAM dendrimers have no effect at comparable concentrations.[30,31] With real cells, hole formation was accompanied by the leakage of cytosolic proteins and by the entry of dendrimer molecules into the cells in an endocytosis-independent patway.[31] The importance of surface charge for the ability to destabilise lipid membranes supports a mode of action (Figure 4.5), involving the displacement of the membrane-stabilising divalent cations found in biological membranes and extremely important for stabilising such membranes by neutralising the anionic groups of the membrane (as also discussed by Chen and Cooper[32]). Dendrimer-mediated displacement may occur through competition by the cationic dendrimer for the divalent cations normally bound by the charged, anionic head groups of the phospolipids or by scavenging the cations in case of anionic dendrimers.

The normal mode of action would be by the competition mechanism, strongly favoured by the more heavily negatively charged prokaryotic membranes and solely involving cationic dendrimers, in a mechanism similar to the action of the well-described cationic antimicrobial peptides, *e.g.* the defensins.[28] However, defensins are typically amphipathic molecules, capable of forming membranes pores through the aggregation into cylindrical "barrels" with hydrophobic outsides that interact

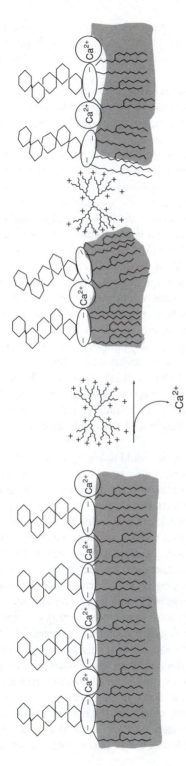

Figure 4.5 *A proposed mechanism for cationic dendrimer-mediated disruption of the surface of a gram negative outer membrane by displacement of Ca^{++} ions by competition for electrostatic binding to anionic groups of the outer membrane lipopolysaccharides (only outer part phospholipids are depicted). The binding of the multicationic dendrimer results in a disruption of the regular structure in which charge neutralisation by Ca^{++} is seen (left)*

with the lipid parts of the bilayer membrane while maintaing a hydrophilic centre that is big enough to allow the diffusion of salt and water.[28] In contrast, dendrimers do not seem to depend on compartmentalised hydrophobic domains for their anti-bacterial effect.

In addition to the positive charge, dendrimers can, however, be designed to contain hydrophobic segments that make them behave like the pore-forming α-helical defensins. For example, certain types of amphiphilic dendrons self-assemble into porous supramolecular columns[33] and constitute highly membrane-active drugs.

The principal mode of action of such dendrimers is thought to equal that of other more conventional bacteriostatic quaternary ammonium compounds, *i.e.* adsorption of the cationic dendrimer onto the negatively charged bacterial cell surface, diffusion through the cell wall and binding to the cytoplasmic membrane, followed by disintegration of the cytoplasmic membrane and subsequent lysis and death of the cell.[34] Dendrimer polycations combine a relatively small size with a high number of charged groups (compared to linear polycationic polymers) and are therefore well-suited to bind strongly and quickly to the bacterial cell surface and may also be able to penetrate the cell wall. The subsequent lysis of the cytoplasmic membrane is also affected more efficiently by cationic dendrimers than by small molecule cations. Finally, dendrimer polycations have a superior ability to complex divalent ions important for the integrity of the membrane.[32,34]

Another group of dendrimers employs surface groups that mimics eukaryotic receptors binding bacteria on cells of various tissue types and in the blood, thereby inhibiting the first step leading to bacterial infection, *viz.* adhesion of the bacterium to host cells/tissues. An important group of such receptors is constituted by eukaryotic cell surface carbohydrates, *e.g.* glycosphingolipids and glycoproteins.[35,36] Yet, another group of antibacterial dendrimers is directly binding to, and thereby neutralising bacterial toxins.

As with the viruses (see above), it should be recognised that efficient inhibition of bacterial adhesion *in vitro* does not equal being able to prevent infection *in vivo*; a substantial problem for *in vivo* uses of such dendrimers is to direct the inhibiting dendrimer drug to the site of infection and to the cells targeted by the bacterium in question. Similar problems of localisation are seen with the other types of antibacterial dendrimers mentioned above; to be efficient *in vivo* they should be directed to the site(s) of infection. Furthermore, antibiotic drugs should be stable and active at physiological salt concentrations and show low eukaryotic cell toxicity.

Antibacterial dendrimers carrying cationic surface functionalities like amines or tetraalkyl ammonium groups include PPI dendrimers functionalised with quaternary alkyl ammonium groups, which were found to be very potent bacteriocides against gram positive as well as gram negative bacteria.[34] Interestingly, tetraalkyl ammonium bromides were found to be more potent antibacterials than the corresponding chlorides.[34] The dendritic bacteriocides were found to have higher activity than other hyperbranched polymers including a polymer containing the same number of quaternary ammonium groups with dodecyl alkyl groups as a G4-PPI dendrimer. Increasing size (and consequently increasing the number of surface cations), on the one hand, should increase antibacterial activity of the dendrimer, but on the other, at the same time renders the drug less cell wall penetrating, but the number of surface cationic groups was found to be the most important parameter.[34] Another parameter was the chain length of the alkyl group on the ammonium groups and here C10 chains were found to be superior to C8 chains and, especially, longer chains (C12, C14 and C16). This may reflect that the antibacterial effect relies on the charge and the hydrophobicity of the dendrimer surface groups, however being too hydrophobic, the dendrimer solubility under physiological conditions begins to decrease.

Antimicrobial peptides have also been used in combination with dendrimers in order to create efficient bacteriocides.[37] Tam and co-workers[37] used peptide-based dendrons (all-lysine dendron containing four or eight terminal groups) terminated with two types of peptides corresponding to antimicrobial consensus peptide sequences. The two peptides (a tetrapeptide and an octapeptide) did not have any antimicrobial activity on their own but became broadly active against bacteria and fungi when conjugated to the tetra- and octameric dendrons, as assayed by a radial diffusion assay. The minimal inhibitory concentration (MIC), which is a standard measure of the activity of antibiotics against bacteria was determined in this assay. A linear polymer of the tetrapeptide was also active but not as broadly (*i.e.* not inhibiting as many different bacterial species) as the dendrimers and some of the linear polymers showed reduced solubility, while others showed increased haematotoxicity compared to the dendritic peptides. Furthermore, the antibacterial activity of the dendrimer peptides was much more protease resistant than the corresponding linear peptides. The dendrimeric tetrapeptides were almost as active as the octapeptides both on tetra- and octavalent dendrimers and retained activity also at physiological salt concentrations. The authors stress that the dendritic peptides are synthesised with fewer steps than the linear peptides in order to obtain the same number of repeating peptide units.

Drugs working as microbial antiadhesins and microbial toxin antagonists may also be based on carbohydrates as recognition molecules on the surface of dendrimers. Dendrimers are useful carriers of carbohydrates as the resulting glycodendrimers have adequate binding affinities even when using simple mono- or oligosaccharides as surface groups, taking advantage of the multivalency/cluster effect (*vide supra*) offered by dendrimer presentation.

Polylysine dendrimers having mannosyl surface groups have been shown to inhibit adhesion of a type 1 fimbriated *Escherichia coli* to horse blood cells in a haemagglutination assay and to model glycoproteins (a binding assay with increased sensitivity as compared to the agglutination assay), making these structures promising as antibacterial agents.[38] *E. coli* binds to host cells by carbohydrate-specific adhesins residing on the fimbriae. The dendrimer drug thereby mimics an artificial cell surface. The best inhibition, 1600 times the inhibition achieved with the free monosaccharide was obtained with a G4-polylysine dendrimer and a neoglycoprotein and efficient inhibition was found to depend on the presence of large numbers of acessible α-mannose moieties bound to the dendrimer surface either through a long alkyl chain or through an aromatic group and the presence on the dendrimer of mannose residues at least 20 nm apart. This corresponds well with the situation for a glycolipid receptor in a cell membrane where the carbohydrate moiety is in close proximity to a hydrophobic aglycon moiety and where the binding of two adhesin molecules situated at the tip of two different fimbria from the same single bacterial would be spatially possible (Figure 4.6).

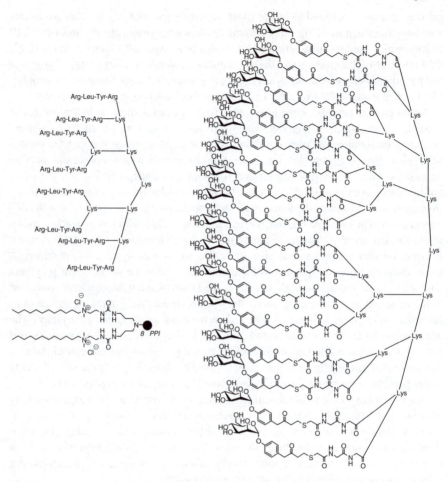

Figure 4.6 *Antibacterial dendrons and dendrimers: top left: antibacterial MAP dendron D_8R4 carrying a tetrapeptide motif[37] in a $(K_2K)_2K$ construct having four lysine α-amino groups and four lysine ε-amino groups available for coupling of an antibacterial peptide. Bottom left: antibacterial dendrimer against E. coli having an alkyl ammonium surface.[34] Right: antibacterial drug against E. coli based on a polylysine dendron having a highly monnosylated surface[38]*

In addition, glycodendrimers have been applied to inhibit the action of soluble, ganglioside reactive bacterial toxins; for example, the inhibition of two different enterotoxins both reacting with GM1, from *E. coli* and *Vibrio cholerae*, respectively were studied using G1- and G2-PPI dendrimers equipped with GM1 tetrasaccharide surface groups coupled to the dendrimer through an aromatic moiety. It was found that this dendrimer inhibited the binding of the toxin as well as toxin sub-units to a fibroblast cell line carrying GM1 on their surface. A subsequent study employed 3,5-di(2-aminoethoxy)benzoic acid-based dendrimers (G1–G3) decorated with lactose, which were found to react with elevated affinity with the GM1 ganglioside-specific cholera toxin B-sub-unit (500 times increase for the octamer compared to monomer) as revealed by a soluble fluorescence assay.[39] Another study on the same bacterial carbohydrate-binding toxins reported the design of a pentavalent glycodendrimer, based on the pentavalent binding of these types of toxins to cell surface glycolipids (GM1 gangliosides), and retaining the geometry of the natural ligands.[40] The scaffold used here was not a traditional dendrimer, but rather a coupling to a pentameric core-linker molecule.[40] In all of these studies, the combination of the surface monosaccharides with an aromatic substituent in itself leads to an increased affinity for the receptor. One could envisage such dendrimers used orally as they should block the interaction of the toxin with epithelial cells within the intestines.[40] A dendrimer designed after similar principles was reported by Nishikawa and co-workers[41] to neutralise shiga toxin produced by *E. coli* and circulating in the blood stream ("Super-Twig" dendrimer). This dendrimer featured trisaccharide globotriaosyl ceramide mimics bound to the surface of a silicon dendrimer with three of these trimers attached to each of the surface silicon atoms. This dendrimer bound directly to shiga toxin thereby inhibiting its binding to its natural glycosyl ceramid receptor and furthermore enhancing phagocytosis of the toxin. In an analysis of dissociation constants with the toxin, it was found that the whole dendrimer series from G1 (four glycosyl groups) to a dendrimer constructed from four G2 dendrons (32 glycosyl groups) has similar binding affinities. There were differences in inhibition potential of the different constructs, a bi-G2 dendron with 18 glycosyl groups showing the highest inhibition towards two different shiga toxin types. This was also tested *in vivo* by simultaneous administration of dendrimer and toxin by the intravenous route and a distance of at least 11 Å between the two trisaccharide clusters was found to be optimal for its function as an inhibitor in circulation as was the presence of at least six trisaccharides; also, clustering of three glycosyls on the same silicon atom was found to be important (Figure 4.7).[41]

As with the antiviral dendrimers, antibacterial dendrimers have mainly been tested *in vitro*, and although a number of studies have shown the ability of antibacterial dendrimers to work well in inhibiting interaction with eukaryotic target structures, adhesion to relevant surfaces and growth, such *in vitro* finding not always correlated completely with the *in vivo* requirements. One important *in vivo* parameter to consider is the immune system, which can aid the dendrimer in removing a pathogen and/or a toxin (see Nishikawa and co-workers[41]), but which may also be adversely reacting by targeting the dendrimer.

4.4 Dendrimers in Antitumour Therapy

Singlet oxygen is strongly damaging to cells and tissues, and *in situ* formation of singlet oxygen can therefore be used to damage unwanted cells and tissues. This can be achieved by irradiation of certain drugs at a defined point in time corresponding to the desired position of the drug.[42] The drug should be non-toxic under non-irradiative conditions (low "dark toxicity"), thus acting as a prodrug when not irradiated (Figure 4.8). Dendrimers containing various photosensitisers for the formation of singlet oxygen in tumour tissues have been described in a few reports only, but constitute a developing field. Dendrimers with 5-aminolevulinic acid photosensitisers at the surface are promising agents for photodynamic therapy (PDT) of tumourogenic

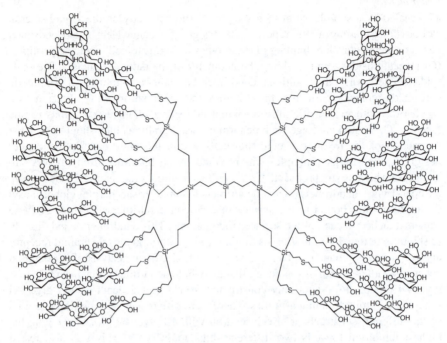

Figure 4.7 *Shiga toxin-neutralising dendrimer ("Super Twig") according to Nishikawa and co-workers.[41] The dendrons terminate in trimer clusters of the toxin-binding Galα1-4Galα1-4Glcβ1 motif and the dendrimer is silicon based*

keratinocytes,[43] and polyaryl ether-based dendrimers derivatised with the photosensitiser protoporphyrin have been evaluated as candidates for the PDT of solid tumours.[44] The protoporphyrin-derivatised dendrimers showed more specific cytotoxicity than protoporphyrin itself, and the dendrimers were more potent upon irradiation compared to protoporphyrin, probably due to an antenna effect of the dendritic wedges. The dendrimers furthermore showed a 140-fold lower dark-toxicity than free protoporhyrin, thus avoiding unspecific cytotoxicity.

Dendrimeric molecules have found use as diagnostic reagents for tumour imaging by magnetic resonance imaging (MRI) and as contrast agents; by varying size and hydrophilicity and by combining with tumour-targetting antibodies, these compounds can be used for a range of specific imaging purposes[45] (see Section 5.1, Chapter 5). By replacing the tumour-binding substance folate on the surface of a polyether dendrimer with the folate-analogue methotrexate, Kono and co-workers[46] could convert a targetting dendrimer into a potentially therapeutic dendrimer targetting folate receptor overexpressing tissue (typically tumours).

Also in treatment of cancer, glycodendrimers constitute an important class of therapeutic molecules, as specific carbohydrate structures may be found on the surface of cancer cells. One such example being the clinically important T-antigen (Gal β1-3 GalNAc), which is characteristic of certain cancer cell types expressing aberrant carbohydrate structures (in particular breast cancer carcinomas). T-antigen has been presented multivalently by coupling to various dendrimers (*e.g.*G1–G4 PAMAM[47,48]).

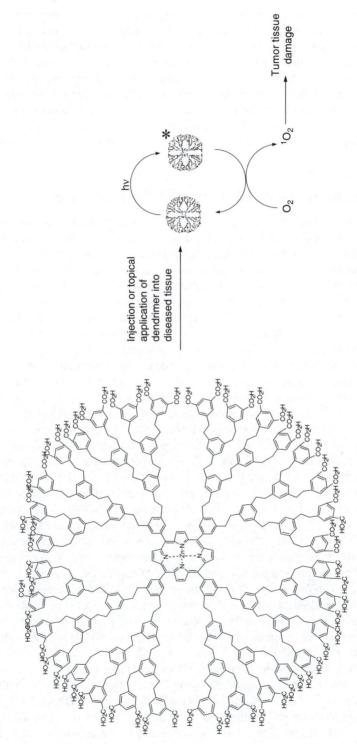

Figure 4.8 *PDT using a dendrimer with a protoporphyrin photosensitiser core. Upon irradiation with light, oxygen surrounding the dendrimer is oxidised to the tissue-damaging singlet oxygen radical (schematic)*

These dendrimers were investigated with the purpose of producing a drug that would interact with carcinoma-derived T-antigen-binding receptors to interfere with carcinoma growth. These types of glycodendrimers reacted in a generation-dependent way with monoclonal antibodies against the T-antigen with higher generations having higher affinities. Considering each carbohydrate unit, the multivalent presentation yielded an approximately 20 times gain in inhibition efficiency towards the antibody, and the "per unit" efficiency did not increase further after the tetramer stage.[47] No clinical results on using these glycodendrimers for treatment of cancers have been presented but their use in breast cancer therapy has been advocated (see review by Roy and Baek[49]).

4.5 Dendrimers in Therapy of Other Diseases

Drug delivery by dendrimers is an important therapeutic application of dendrimers and is described in Chapter 3, which also deals with dendrimers as non-viral DNA transfection agents, also having obvious therapeutic applications (gene therapy, antisense treatment of cancer and virus infections). The use of dendritic drugs to combat bacterial and viral infections is described in Sections 4.1 and 4.2. Here will be mentioned just a few examples showing that dendrimers may also be used directly as drugs for therapeutic purposes. These applications are based on the key features of dendrimers, *i.e.* their high degree of molecular definition and their derivatisability meaning that they can be tailored to suit specific purposes. For example, some of these drugs work by a targetting action of the dendrimer, others utilise the general ability of a range of dendrimers to interact with or even penetrate biological membranes and some of them rely on the basic ability of dendrimers to amplify the effect of a given drug by its multimeric presentation. Also, dendrimers may simply serve to control solubility and retention in circulation *in vivo* of the drug in question.

Interesting examples of biologically active peptides for which increased biological activity has been achieved by a dendritic design are peptides derived from the neural cell adhesion molecule (NCAM); both an 11-mer (C3), 12-mer (P2) and a 15-mer (FGL) have been described as having increased biological activity when presented in a tetrameric dendron based on a multiple antigenic peptide (MAP) lysine core (*vide infra*, Section 4.6).[50] These peptides have strong effects on neuronal plasticity and regeneration and especially the FGL-peptide which works as an agonist for the fibroblast growth receptor seems to have general, neuroprotective effects in models of neuronal cell death[51] and promotes memory functions in a rat model.[52]

Although no molecular mechanisms have been described to account for the increased biological effects of these dendritic peptides, the effect is probably due to the increased affinities observed for the binding of these peptides to their receptors which may rely on the synergistic dendritic effect described elsewhere in this volume.

Flexibility was considered a key feature of dendrimers designed to react with RNA in a study describing the interaction of triethanolamin core cationic polyamine dendrimers with *Candida albicans* ribozyme RNA.[53] This is due to the multiple structures and sizes of RNA encountered in nature (in contrast to DNA). These dendrimers could block the enzymatic reactivity of the ribozyme (RNA oligonucleotide cleavage and splicing reactions). Inhibition efficiency increased with the generation

(G2–G4) and was dependent on the presence of accessible, primary, tertiary or quaternary amines on the dendrimer surface and attributable to electrostatic interactions. Direct binding between the dendrimers and the RNA could be demonstrated. Binding to RNA and blockage of a viral RNA-binding peptide was also demonstrated by Zhao and co-workers[54] using G3 PAMAM and synthetic RNA representing the small transacting responsive element, which is essential for HIV-1 replication. While these studies employed extracted or synthetic RNA, any *in vivo* application of such a method for blocking specific RNA-species would demand the crossing of cell membranes. This, as described above, is often achievable with dendrimers and might be combined with peptidic transmembrane transporters as *e.g.* the TAT-1 peptide to transport molecular beacon oligonucleotides into cells for subsequent binding and labelling of specific RNA species.[55]

The possibility of targetting dendrimers to certain cells and/or tissues expressing disease-related molecular motifs often used for diagnostic purposes (see Chapter 5) can obviously also be utilised to increase the efficiency of treatment with drugs either carried by a dendrimer (drug delivery, see Chapter 3) or constituted by the dendrimer itself. The ability of dendrimers to cross cell membranes can be used to target intracellular components. The biggest promise for therapeutic uses of dendrimers may be within the cancer field where numerous examples of targetting tumours for diagnostic purposes have been described and where it is often possible to define a cancer-specific, accessible cell surface component that can serve as a target.

4.6   Dendrimer-Based Vaccines

It is beyond the scope of this text to describe the immune system and its stimulation by vaccines in any detail, but a few terms and basic principles will be explained here to introduce the reader to the subject area.

Vaccines exploit the ability of the vertebrate immune system to mount an adaptive response against specific pathogens entering the body, thereby protecting the host against the given pathogen. Vaccines can be defined as non-pathogenic mimics of pathogens, used to induce immunity against the pathogen, by injecting or administering the vaccine into the host (vaccination). It is a hallmark of the immune system that it reacts towards and is induced by non-self structures; furthermore, it is stimulated by certain structures perceived as "dangerous" by the immune system.[56] Last but not least, an important component of immunity is immunological "memory" ensuring that specific immunity is quickly and efficiently reactivated upon infection with the specific pathogen to which the vaccine was directed. Importantly, in order for a vaccine to be able to efficiently provoke a desired host immune response, it has to be mixed with substances that facilitate this process (adjuvants) and/or the vaccine has to be modified in order to increase the immunostimulating properties (the so-called immunogenicity) of the pathogen mimic. *Immunogens* are the components of a vaccine towards which the desired immune response is to be directed; an *antigen* is a substance that will bind an antibody in an antigen-specific way. Some antigens will be immunogens by themselves, but this is not always the case. An *antibody* is the active component of humoral immunity, and is constituted by the class of soluble proteins known as *immunoglobulins*.

The immunity achieved by vaccination is mediated by humoral factors (immunoglobulins) and/or cellular actors (T- and B-cells, macrophages, *etc.*) and is aided by the complement system and a complicated system of intercellular mediator proteins (cytokines) controlling both immune and inflammatory responses. Immunoglobulins are produced by *B-cells*, while cellular immunity and factors stimulating B-cells are provided by *T-cells*. *Inflammation* is the immediate response of the tissue towards infection (*i.e.* invasion of microbial pathogens) or injury and is characterised by accumulation of fluid and activated cells at the site of the inflammation. Inflammation is thus indicative of tissue destruction.

Most of these reactions involve the binding of soluble components to cell surfaces and to cell surface receptors. It is important to realise that the immune system, through a large number of common mediator substances is interacting closely with other cell-based, responsive systems like the complement system, coagulation system, inflammation cascades, *etc.*, each of which can have enormous consequences for the host. For vaccine development, it is therefore very important to be able to control unwanted side-effects stemming from aberrant activation of these other host defence systems. In addition, some types of immune responses, especially those risking to induce autoimmunity, are also undesirable.

Vaccination remains the most cost-effective way of preventing or even treating infections. Furthermore, vaccines are finding use as cancer therapeutics and for deliberately creating autoreactivity in order to control endogenous, undesired immune and inflammatory reactions.[57,58] An efficient vaccine should be highly specific and highly active *i.e.* inducing the immune reponses needed quickly, efficiently (leading to protective host reactivity) with the creation of immunological memory and without adverse side-reactions. Complete protection against an invading pathogen normally depends on both T- and B-cell immunity to be induced (Figure 4.9).

It is often desirable to make vaccines highly specific to avoid problems of autoreactivity and other side-effects. This can be done by immunising with a small part (the so-called sub-unit) of the pathogen, or even a small part of a specific component of the pathogen, for example, a peptide corresponding to a part of a microbial protein. However, it is well established that small molecular weight substances (*e.g.* peptides) are not very immunogenic *i.e.* no or a weak immune response (including antibody formation) is induced upon their injection into a recipient host. This problem can be overcome by increasing the molecular weight of the substance in question either by polymerisation (as *e.g.* "peptomers"[59] or by the free radical polymerisation method of Jackson and co-workers[60] using acryloyl peptides), or by coupling it to a multifunctional, high molecular weight carrier.

As will be seen below, dendrimers have properties of multivalency, size and structural definition that allow them to be used as building blocks, scaffolds or carriers for creating efficient vaccine components (immunogens) that will induce a desired and predetermined type of immunity in the vaccinated host. Traditionally, *carrier molecules* have been non-self (foreign) proteins used to bind small, non-immunogenic antigens (*haptens*) to make them immunogenic. Examples include naturally derived proteins as *e.g.* ovalbumin, keyhole limpet, haemocyanin, various toxoids, *etc.*[61] In such a construct, immunogenicity is thought to arise due to the *multimeric presentation* of the antigen, due to the presence in the carrier protein of T-cell-inducing epitopes and due

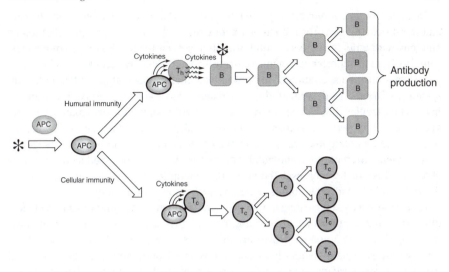

Figure 4.9 *Overview of the activation of the immune system upon contact with a "foreign" substance, in this case a dendrimer carrying antigenic moieties (asterisk). First, the antigen is taken up, processed and presented on the surface to the T-cells of the immune system by antigen presenting cells (APC). T-cells having binding activity against the specific antigen in question recognise the antigen presented on the APC surface in complex with the so-called MHC molecule and thereby becomes stimulated to proliferate and to secrete cytokines. These cytokines stimulate B-cells expressing surface bound antibody molecules with specificity towards the antigen to proliferate and secrete antibody (immunoglobulin). This results in an antigen-specific propagation of T-cells (Th (helper cells) and sometimes Tc (cytotoxic) cells) and B-cells producing antibodies. Cell–cell signals are provided by mediator molecules called cytokines*

to the size of the complex, crossing the lower antigen size limit of the immune system, which is around 2–3 kD.[56] Also, the carrier protein will supply immunogenic peptides able to stimulate T-cells that will help in achieving a full-blown immune response (peptides binding the major histocompatibility complex (MHC) which has a restricted specificity). Finally, it is conceivable that the coupling of conformationally undefined antigens as *e.g.* most peptides stabilises their conformation slightly resulting in a more appropiate immune response. These methods are largely empirical and it is not possible to predict optimal carriers, conjugation chemistry and carrier–peptide coupling ratios for desired immune responses. Also, the conjugates are not easily analysable and have unknown coupling ratios, unkown coupling orientation of the peptide antigen and they contain impurities consisting of poly- and oligomers of the peptide and the carrier protein. Finally, naturally derived carrier proteins are not fully chemically characterised or standardisable and may be expensive.

For the preparation of highly defined, reproducible immunogens, *e.g.* for human vaccine uses, other types of carriers are highly desirable, and in this respect dendrimers have emerged being useful as they can act as multivalent and well-defined carriers for antigenic substances by coupling of antigen molecules to the surface functional groups of the dendrimer.

Dendrimers for preparation of efficient, fully synthetic immunogens. Dendrimers and the processes used for their manufacture allow this multimeric presentation of antigens with a degree of definition normally reserved for much smaller molecules. This is of great importance when exploring ways to decrease unwanted side-effect as *e.g.* uncontrolled activation of inflammation, and the coagulation and complement systems. Also and as described elsewhere, dendrimers are very versatile and easily handled molecules that can be derivatised in a highly controllable manner. Synthetic vaccines, as opposed to vaccines based on naturally derived molecules or even dead or attenuated whole pathogens are comparatively stable and robust and thus can be used under different environmental conditions and can be produced relatively cheaply. This greatly increases the potential general benefit of vaccines as some of the most widespread infectious diseases occur in third world countries.[56]

Dendrimers themselves are generally considered to be non-immunogenic, which is obviously important for their *in vivo* uses *e.g.* as drugs and for drug delivery applications (see Chapters 3 and 4). However, although the induction of adverse host reactions by *in vivo* administration of such substances does not seem to have been explored in any systematic way, some data are available to indicate that this is not always a negligible problem (see Chapter 2). For example, the complement-activating ability of DNA–dendrimer complexes has been reported and is clearly a disadvantage when considering using dendrimer-mediated DNA transfection *in vivo e.g.* for therapeutic purposes.[62] This, by contrast might be exploited, in a vaccine context, to broaden the adjuvanticity of certain DNA sequences (see discussion below). An *adjuvant* is a substance that will augment the immune response towards an antigen upon coadministration into a host of the adjuvant with the antigen. Most adjuvants function by stimulating immune-related cells unspecifically, and/or by restricting the release of antigens, prolonging local exposure to the antigen (depot effect).

In most cases, dendrimers may be considered immunological inert scaffolds that need to be decorated with antigen (normally in a multimeric fashion) and combined with adjuvant in order to function as a vaccine immunogen.

The MAP system. Most work with molecularly defined immunogens has been done with the peptide dendron MAP construct. The MAP system was pioneered by Tam and co-workers[63–66] and is by far the most succesful dendrimer type used for vaccine and immunisation purposes. MAP structures have been used in a large number of studies for producing peptide-specific antibodies[64] and are also being developed for vaccine use, MAP-based malaria vaccine being in phase I human trials.[67–69] MAPs have also been used succesfully for the efficient presentation of antigens for the detection of antibodies, at the same time increasing the antigen concentration and density on a surface and supporting the structure of such antigens allowing the detection of low-affinity antibodies (see also Chapter 5 for dendrimer amplification of bioassays).[70]

The MAP construct has the advantage of being conceptually and practically simple as the dendron carrier (the MAP "core") is composed solely of amino acids (typically lysine) and can be synthesised by conventional solid-phase peptide synthesis, and as the antigenic peptides coupled to the carrier can be synthesised directly on the carrier by conventional chemical methods. The method, however also allows convergent synthesis, also called segment coupling of antigenic moieties, which may include substances other than peptides (Figure 4.10). The design aims to achieve the highest possible degree of multimericity in the smallest possible space (high-density multimer), which is claimed to ensure optimal immunogenicity and furthermore stabilises the structure of the attached peptides, assures absence of interference from non-relevant epitopes in the carrier and provides a sufficiently high molecular weight to ensure immunogenicity (typically >10 kD) (see reviews by Haro and Gomara,[71] Veprek and Jezek,[72] Tam,[73] Nardin and co-workers[74] and Sadler and Tam[64]). It is evident that the most simple form of the MAP construct does not provide MHC-restricted T-cell epitopes to overcome MHC-restriction and also needs adjuvant to function optimally. Also in some cases, MAPs are not readily soluble in aqueous solvents, forming undefined intermolecular aggregates probably accompanied by intramolecular collapse of peptide conformations. On the other hand, the aggregate formation/limited solubility may aid in increasing the immunogenicity of the compound, maybe even partly acting as an adjuvant (*vide infra*) and allow its survival and slow "release" *in vivo*, but on the expense of the definability. A central feature of MAPs is that they should induce immunity against the cognate protein from which the peptides in the MAP have been derived, but this is not always the case which illustrates that the antigenic peptide part of MAPs does not always attain a structure similar to the native structure of the peptide in the cognate protein,[70] although the stabilisation of native peptide structure has been directly shown by spectroscopic methods with some peptides (*e.g.* an amphipathic peptide in a tetravalent MAP[75]).

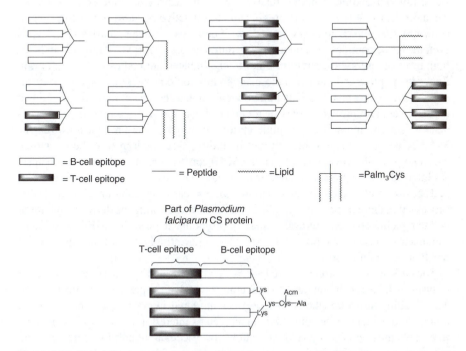

Figure 4.10 *Top: different arrangements on the MAP structure to afford the proper vaccine motif. Bottom: the T1B4 MAP vaccine structure developed by Moreno and co-workers[67]*

To ameliorate some of these drawbacks, a number of different variants of the basic MAP construct have been reported, including MAPs comprising defined mixtures of B- and T-cell epitopes,[76] either synthesised by stepwise peptide synthesis on the branches of the MAP or by segment coupling of peptide fragments by various methods[64] and also including lipid-containing MAPs (simple lipids, lipid core peptides (LCPs),[77,78] tripalmitate structures[79]), the lipid serving as a built-in adjuvant (*vide infra*). Especially, MAPs with built-in adjuvanticity are interesting as they combine a high level of definition with an efficient mode of action and as only one type of traditional adjuvant is allowed for use in humans.

To circumvent MHC-restriction of T-cell activation broadly reactive ("universal") T-cell epitopes may be included together with the antigenic peptide of interest (see *e.g.* Tam[73]). This can be done in numerous ways (Figure 4.10), including attaching the B- and the T-cell epitope in tandem. Cavenaugh and co-workers[80] observed that MAPs restored the immunogenicity of a lysozyme T-cell epitopic peptide to the level of the same peptide when administered as part of a whole protein (lysozyme) using fluorescein as the hapten and measuring antifluorescein antibody responses. The response was 300-fold observed after immunisation with the monomeric T-cell peptide–fluorescein conjugate, and the response towards fluorescein was almost absent using a Gly_{18}-coupled tetrameric MAP construct. The conclusion is that MAP multimerically presented T-cell stimulating peptides are far more potent than equimolar concentrations of linear monomers of the same peptide.

There are also some indications that MAPs owing to their structure have an intrinsic ability to overcome at least some types of MHC-restriction[73] and even sometimes can activate T-cells without the presence of T-cell active peptides in the construct as shown by Olszewska and co-workers[81] with peptides from the measles virus fusion protein. These peptides, when incorporated into an octameric MAP not only induced high titres of antibodies, but also, importantly, induced antibodies with a high affinity for the peptide and the virus; these antibodies could protect mice against measles virus-induced encephalitis. This immunogenicity was comparable to that of a linear construct containing two copies of a universal T-helper cell epitope from measles virus F protein and a B-cell peptide epitope from the same protein, however, the MAP construct achieved this without including T-cell epitopes in the construct. T-cell help was however engaged by the MAP construct by an unknown mechanism (as inferred from lymphocyte proliferation data). Using cancer-related peptides, Ota and co-workers[82] showed that certain MAPs carrying tumour antigens were processed in antigen-presenting cells in the same way as antigens derived from intracellular pathogens (*e.g.* viruses), thereby providing a powerful MHC I-restricted immune response, including cytotoxic T-cells; interestingly this was achieved with a MAP not containing lipid moieties (*vide infra*).

The basic MAP construct is a wedge-like dendron formed by successive generations of lysine residues acylating the α- and ε amino groups of the preceding lysine residues. The resulting structure (the MAP core) has an equal number of α- and ε-primary amino groups that can be coupled to an antigen of interest, typically a synthetic peptide. The most preferred MAP-structures for vaccination purposes are tetra- or octameric as higher generations tend to cause problems during synthesis (incomplete acylations), solubility problems of the final compounds with some antigenic peptides and do not improve immunogenicity.[72,73] The basic strategy employs stepwise solid-phase synthesis on the resin-bound MAP core for the attachment of antigenic peptides,[63] however by using orthogonal protection strategies and/or chemoselective coupling methods, different types of unprotected peptides may be coupled in a controlled fashion to the same MAP carrier.[72,83] This adds to the versatility of the method and is also helpful with regards to the synthesis, as it is well known that stepwise solid-phase synthesis is quite prone to interference of acylations by aggregation of the growing peptide chains during synthesis, especially at high-synthesis densities,[84] and as unprotected segment coupling allows a purification step to be applied before coupling of the antigenic moieties to the MAP core, facilitating the final purification of the MAP product. Examples of such chemical methods include coupling of thiol nucleophiles to haloacetylated amines; this could be a cysteine-containing peptide or a peptide thiocarboxylic acid reacting with an α, ε bromoacetylated MAP core, resulting in thioether and thioester bound peptides (however, it should be noted that thioesters are unstable at pH 7 and above). As a cysteine residue can be placed anywhere in a synthetic peptides, this allows for freedom in selecting the orientation of the peptide. Bromoacetylation might be controlled by using selective protecting groups on the α and ε amino groups, respectively. Another useful group of reactions is the condensation of aldehydes (not normally found in peptides) with weak bases under acidic conditions that inhibit the reactivity of side-chain nucleophiles with the weak base; useful weak bases are hydroxylamine and substituted hydrazine that produce oxime and substituted hydrazone, respectively, with carbonyl groups. For example, the MAP core may be decorated with aldehydes by mild oxidation of serine, threonine or cysteine coupled to the N-termini of the MAP core and then reacted with aminoxyacetylated peptide to form the oxime or with peptide acylhydrazine to form the substituted hydrazone (see Tam[73] and Tam and Spetzler[83] for a good overview of these chemistries).

A large number of lysine derivatives with orthogonal protection on the two amino groups are commercially available for constructing MAPs with two or more different types of peptides. A common goal is to include both a B-cell peptide epitope, stimulating antibody development, and a T-cell epitope, stimulating cellular immunity, as optimal immunity is normally dependent on both types of activities being present. Spacers may also be included anywhere in the construct. Finally, cyclic peptides, showing a further increased structure stabilisation can be incorporated into a MAP structure[85] as used by de Oliveira and co-workers[86] for preparing a well-defined four-valent foot-and-mouth disease virus lipopeptide MAP analysable by HPLC and mass spectrometry; although the intention was to produce the four-copy construct only, the three-copy construct was obtained which was thought to be due to steric hindrance in the segment coupling reaction. It has been noted by others that some commercial MAP core resins do not yield the valency that they are supposed to.[80]

By far, most of the reported immunisations with MAP constructs have been performed with traditional adjuvants as *e.g.* in the work by Moreno and co-workers[67] in which aluminium hydroxide, Freund's adjuvant and a saponin adjuvant (QS-21) were tested for the ability to induce antibodies together with a MAP structure containing *Plasmodium falciparum* T- and B-cell stimulatory peptides in different species of monkeys and in mice using subcutaneous administration. This MAP construct was claimed to be homogenous as judged by HPLC.[67] Interestingly, an octameric feline immunodeficiency virus (FIV) MAP administered with Freund's adjuvant led to high levels of neutralising antibodies, while administration with QS-21, giving rise to antibodies and cytotoxic cellular responses to the immunising peptide did not induce neutralising activity.[87] Thus, also for MAPs the formulation and administration regimes are of prime importance for the response obtained.

The basic MAP construct has been developed further to comprise moieties with adjuvant activity (self-adjuvanting vaccine delivery system[78]). This allows immunisation with MAPs without adjuvants, dramatically broadening the application area and facilitating their use. Such MAPs constitute truly fully synthetic immunogens with specific adjuvant characteristics and minimised adverse effects. This is accomplished by coupling the MAP dendron with a lipid-containing moiety, *e.g.* the rather complicated tripalmitate-*S*-glyceryl cysteine structure mimicking gram negative bacterial membrane components first described by Deres and co-workers[79] and simple fatty acid moieties as palmitic or myristic acids[88,65] or amino acids with lipidic side chains (LCPs[77]). These structures dramatically enhance the immune-inducing activity of the peptidic constructs probably by enhancing contacts with membranes of immune cells and thereby both optimising exposure of the peptide antigens to the immune cells and at the same time activating the cells to secrete activating substances (cytokines). Lipid and/or amphiphilic substances are found in most traditional adjuvants and most compounds with general immune-stimulating properties (lipopolysaccharides, phorbol myristyl acetate, *etc.*). In addition, lipids may promote the uptake into antigen-presenting cells (APC) through a pathway leading to the so-called MHC I-presentation, which is a prerequisite for induction of the so-called cytotoxic T-cells that are crucial for the combat of intracellular infections including viral infections[82] (*vide supra*).

Examples on the use of lipoaminoacid-containing MAPs (LCPs) include the tetrameric MAP with 2×2 peptides corresponding to different regions of a streptococcal membrane protein, elongated C-terminal with a pentapeptide containing three lipidic (octyl) side chains.[78,89] By subcutaneous immunisation without added adjuvant this construct led to antibody responses and protection levels similar to immunisation in the presence of Freund's complete adjuvant (the most powerful adjuvant known). LCPs have also been shown to be able to induce cytotoxic T-cells using an ovalbumin model peptide and were shown to be able to slow down the development of tumours in a mouse model, provided an additional adjuvant (alum) was used.[90] Finally, lipidated MAPs have potential to be used for mucosal immunisation as they are supposed to be ideally suited to penetrate mucosal membranes, an example being a tetravalent HVI-peptide lipidated MAP (lipid: tripalmitate structure) that was able to induce IgA at the mucosa and serum IgG responses by oral delivery, and cytotoxic MHC class I-dependent T-cell responses by intragastric delivery.[91]

Iglesias and co-workers[92] investigated dimers of B- and T-cell epitope MAP dendrimers as well as such dendrimers coupled to carrier proteins where the B-cell epitopes originated from the highly variable V3 loop of HIV-1 gp120 and the T-cell epitope originated from tetanus toxoid. Dimeric MAPs were prepared by disulfide coupling between C-terminal cysteines in the MAPs. It could be shown that the immune reactivity resulting from immunisations with the MAP–protein conjugates had a broader reactivity to various heterologous V3 peptides than obtained by immunisation with pure MAP constructs while the immunogenicity was similar. Hepatitis

B surface antigen was used as the carrier protein. Cruz and co-workers[93] also showed increased cross-reactivity with the same peptides using HBsAg coupled tetrameric MAPs also containing a tetanus toxoid peptide T-cell epitope (tandem peptide chimera); however in this case, the coupling to the carrier protein also significantly increased the immunogenicity.

Baek and Roy[94] explicitly claim that multimeric carbohydrate moieties as presented in glycodendrimers are non-immunogenic, and it is well known that carbohydrates *per se* normally show low immunogenicity.[56] However, many biologically relevant carbohydrates are interesting targets for protective immune responses, including cancer-[57] and virus-specific carbohydrates and bacterial cell surface carbohydrates and therefore methods to increase the immunogenicity of carbohydrates are very welcome. An example of a MAP construct for glycoimmunogens is the Tn-antigenic dendrimer studied by Bay and co-workers[95] where a tetrameric core structure is derivatised with the Tn-antigen and with a Th-cell stimulatory peptide and was shown to react with monoclonal antibodies against Tn (Figure 4.11). A different version of this, also containing trimeric Tn-building blocks was later shown to be immunogenic,[96] and useful for active immunisation against colon carcinomas in BALB/c mice, using alum as the adjuvant. As the mono-Tn analogue was less efficient than the tri-Tn analogue and as a linear analogue containing two tri-Tn moieties was also less efficient, it was concluded that the precise spatial arrangement and clustering of the Tn-epitope was very important for the immunogenicity. G5-PAMAM (Starburst™) dendrimers have been applied as carriers of the Tn-antigen and the resulting glycoconjugates were tested as vaccine candidates in comparison with a carrier protein (bovine serum albumin) conjugated to a monomer, dimer or trimer of the Tn-antigen. It was found that the

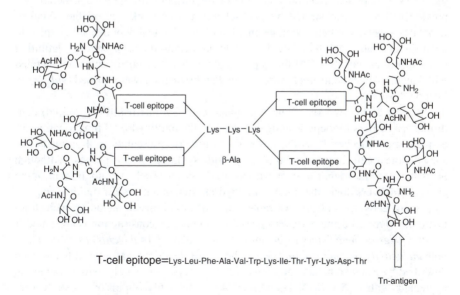

T-cell epitope=Lys-Leu-Phe-Ala-Val-Trp-Lys-Ile-Thr-Tyr-Lys-Asp-Thr

Tn-antigen

Figure 4.11 *A MAP-based vaccine comprising peptides representing T-cell stimulating epitopes in addition to trimeric Tn-epitopes (Galactosyl-threonine)*[95]

Tn-antigen–dendrimer conjugates elicited no antibody response, and hence no immunogenicity, whereas Tn-antigen conjugated with a carrier protein or lipopeptide gave rise to antibody responses. The Tn-dimer lipopeptide conjugate also gave rise to IgG antibodies.[97] G2-PAMAM GlcNAc$_8$ was shown, upon ip or im administration to mice to result in increases in T-cell activity and an enhanced natural killer cell activity, and also reduced tumour growth of previously inoculated melanoma cells in mice.[98] GlcNac is not exposed in a terminal, non-reducing position in cell-surface glycoconjugates, but certain cancer cells express glycoconjugates terminated with this monosaccharide. It was shown that an enhanced host-immune reactivity, including cellular and cytokine factors against the innoculated melanoma cells was induced upon vaccination. This constitutes an example of therapeutic vaccination against cancer inducing an antitumour response.

Other dendrimer-based methods for increasing immunogenicity. Mihov and co-workers[99] investigated the so-called shape-persistent multiple peptide conjugates composed of polyphenylene dendrimers (confusingly abbreviated PPD, a common abbreviation for "Purified Protein Derivative" which is a well-known mycobacterial protein widely used as a carrier protein) onto which polylysine was grafted using various chemistries; these dendrimers were claimed to be optimal for supporting the secondary structure of antigenic peptide attached to their surface. The use of these carriers for immunisation was however not reported.

Other peptide carrier systems, which are not dendrimeric *per se* but becomes a dendrigraft structure upon derivatisation with peptides is the peptide carrier by Heegaard and co-workers[88] and the sequential oligopeptide carriers of Tsikaris and co-workers,[100] in which the attachment points for the peptide branches are designed to space the attached peptides in an optimal fashion supporting structural trends in the attached peptides. This phenomenon of organisationally induced structure has previously been demonstrated by Tuchscherer and co-workers[101] in the so-called template-assisted synthetic peptides in which four identical peptides are coupled to a tetrafunctional, cyclic template, leading to an increased conformational definition of the peptides, compared to the peptides alone. This has also been demonstrated with leucine zipper dendrimers where coiled coil structures were formed (*vide infra*, Section 4.5).[102]

McGeary and co-workers[103] have prepared carbohydrate-based (glycolipid) dendrimers as carriers for peptide antigens, utilising the multihydroxyl functionalities of a single monosaccharide as the basis for multimeric presentation of antigens and showing the applicability of solid-phase peptide synthesis for this purpose. Although no actual peptide–dendrimer constructs were synthesised and no immunisation experiments were done, the use of carbohydrate functionalities for multimeric presentation of antigens is clearly warranted, and the possibility of including such structures into glycodendrimers (*vide supra*) for immunogen construction is presented.

Other uses of dendrimers to increase immune activity (adjuvants) or to decrease immune activity (immunosuppression). The use of dendrimers as adjuvants has been described by Rajananthanan and co-workers[104] comparing two glycolipid-containing aggregates with a G5-PAMAM dendrimer. As such, the glycolipids of this study were expected to be much more amphiphilic than the dendrimer, and moreover they were meticulously formulated with various other components to prepare multimolecular complexes with non-covalently entrapped antigen *ad modum* Iscoms.[105] However, when testing immunogenicity in mice with a standard protein antigen (ovalbumin), mixing antigen and dendrimer increased the immune response above that seen when administering the antigen alone, reaching titres in the 10^5 range being 10 times the titres reached with the antigen alone. Wright claims that G3-PAMAM and other mid-generation dendrimers can be used as adjuvants for vaccine purposes when used in a dilution ensuring low toxicity.[106] Generally, adjuvanticity as measured by antibody titres following immunisation of mice with an influenza antigen adjuvanted with the dendrimer, increased with increasing generation of PAMAM from G0 to G6.

Baird and co-workers[107] investigated various dendrimer (G1, G2 and G3-PPI dendrimers) constructs for presentation of a hapten (dinitrophenyl, DNP) to mast cell surface bound IgE with the purpose of inhibiting the interaction of these IgE molecules with the monovalent hapten, without triggering the release of allergic mediators (Figure 4.12). This is thought to occur by intramolecular cross-linking of the IgE molecules (taking advantage of the dendritic (cluster) effect to achieve tight binding), while intermolecular cross-linking of IgE molecules leads to mast cell activation and should be avoided. The G1 DNP-decorated PPI (tetramer) showed the highest inhibition followed by G2 (octamer), while G3 (16 mer) led to activation of the mast cells (degranulation); probably the bigger size of the DNP G3 PPI enabled

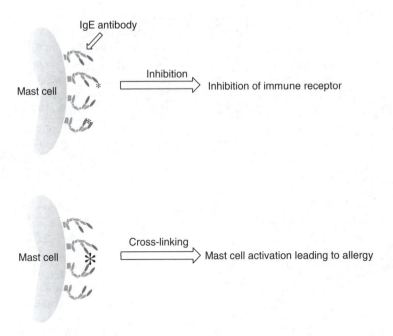

Figure 4.12 *Mast cell surface IgE can be inhibited from binding and being cross-linked by allergens if a synthetic, non-cross-linking agent, e.g. a moderately sized dendrimer (asterisks) carrying the relevant antigenic motif is added (top). However, if the dendrimer inhibitor is too big (bottom), it will cross-link the IgE molecules and instead initiate an undesired mast cell activation leading to allergic reactions. The IgE is bound to the cell surface by Fcε receptors[107]*

it to create intermolecular cross-links between mast cell surface IgE leading to the undesired activation of the cell. Also, Shaunak and co-workers[108] describe the ability of G3.5 PAMAM decorated with either glucosamine or glucosamine 6-sulfate (Figure 4.13) to modulate (inhibit) biological responses; the glucosamine polymer inhibited Toll-like receptor 4-mediated lipopolysaccharide-induced synthesis of proinflammatory cytokines and the glucosamine 6-sulfate dendrimer inhibited angiogenesis and together they prevented scar tissue formation.

To summarise, dendrimers generally have been surprisingly little studied as vaccine components and/or immunogen carriers, which is surprising considering the ideal qualities of dendrimers for this purpose (multimericity, derivatisability, high definition and high molecular weight). The exception to this is the MAP-type dendrimers which have emerged from the peptide field[63] and which have been specifically developed for presenting small peptide antigens to the immune system. MAP cores have later been demonstrated to be good carriers for non-peptidic haptens too (*e.g.* carbohydrates[97]) and to be able to accommodate several different types of peptides at the same time, typically B-cell stimulatory peptides together with T-cell

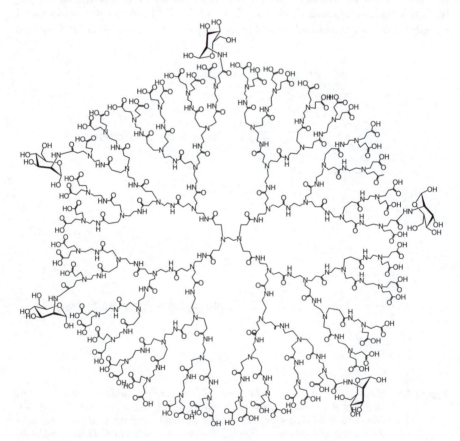

Figure 4.13 *Antiinflammatory dendrimer, based on a G4.5-PAMAM dendrimer partially modified at the surface with glucosamine[108]*

stimulatory peptides. Furthermore, convergent strategies for the synthesis of MAP constructs have been developed[73] as has the possiblity of introducing lipid moieties[77] as built-in adjuvants.

Dendrimer–DNA complexes have been shown to activate complement as an adverse effect of using dendrimers as transfection reagents *in vivo* thereby limiting their therapeutic potential (see Chapter 2).[62] It could be envisioned that this effect could be exploited to potentiate the recently described adjuvant effect of umethylated, CpG-containing oligodeoxynucleotides (Krieg[109]) by administering these compounds together with a dendrimer; this could potentially lead to an increased effect towards cellular pathogens (complement-mediated lysis).

4.7 Dendrimer Interactions with Proteins. Solublilisation of Protein Aggregates

Dendrimers are large solutes that given an adequate level of polarity and charge are soluble in water and exhibit complex interactions with other aqueous soluble biological molecules like lipid bilayer membranes (see above) and peptides and proteins. In addition to specific interactions with other solubilised biomolecules, some dendrimers exhibit major effects on the properties of the solvent often in a detergent-like fashion as will be described below for PAMAM and PPI dendrimers at certain conditions of pH. Such general solvent effects may also play a role in the specific interactions of dendrimers with proteins, and especially lipid membranes (see also Chapter 2).

The litterature on the general interactions between dendrimers and aqueous solutions of proteins is rather limited, but illustrative examples are the reports by Supattapone and co-workers[110,111] and Solassol and co-workers[112] on the effect of dendrimers on the formation and stability of insoluble aggregates of the prion protein. Also, Ottaviani and co-workers[113] reported studies on the interaction of PAMAM with model proteins.

Prion protein (PrP) aggregates are interesting examples of the consequences of protein misfolding. Such aggregates are hallmarks of the so-called prion diseases, which include Creutzfeldt-Jakob's disease, mad cow disease (bovine spongiform encephalopathy) and a range of other neurodegenerative, fatal diseases. The very insoluble aggregates are found in the brains of affected individuals where they precipitate and are associated with spongiosis, inflammation and neuronal death. PrP aggregates are composed of an abnormal conformer of the otherwise innoxious prion protein, and can only be solubilised in aqueous buffers containing both a detergent and a chaotropic denaturant (as 6 M guanidinium chloride). Misfolded, aggregated prion protein is identified by its high protease resistance (Figure 4.14). Similar aggregates, although involving other polypeptides are found in other protein misfolding diseases like Alzheimer's disease, Huntington's disease and others.

It is therefore of considerable interest also from a medical point of view that cationic dendrimers were found to be able to solubilise PrP aggregates as first demonstrated by Supattapone and co-workers.[110,111] This was a chance finding based on the fact that the transfection reagent Superfect (TM, Quiagen), which is a heat-fractured dendrimer,

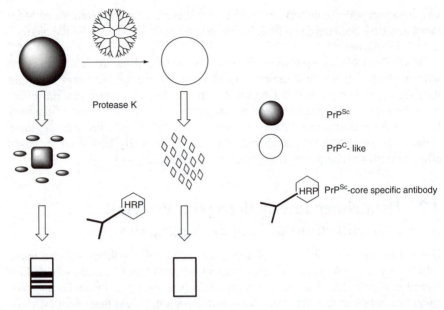

Figure 4.14 *Prion proteins, dendrimers and their analysis. Prion proteins exist in a normal, non-pathogenic and non-contagious form (PrP^C) as well as in a rare, misfolded form (PrP^Sc) that causes disease and can transmit from one individual to another. A hallmark of PrP^Sc is its protease-resistant aggregates. This allows PrP^Sc to be detected by antiprion antibodies after protease treatment in contrast to PrP^C which is completely degraded to non-immunoreactive fragments by the protease. Supattapone and co-workers[110,111] discovered that certain dendrimers could render PrP^Sc protease sensitive (i.e. PrP^C like)*

used to transfect neuroblastoma cell lines with prion DNA resulted in the absence of protease-resistant PrP aggregates, while when using other types of transfection reagents the cells sucessfully synthesised such molecules.[110] This lead to the investigation of 14 different polyamines for their ability to perturb the production of aggregated PrP by chronically infected neuronal cell lines in cell culture. The dendrimers included PPI, PEI and PAMAM dendrimers, and higher generation (>G3) dendrimers were shown to be the most efficient and the effect was correlated to the number of surface amino groups. PAMAM dendrimers having surface hydroxy groups (PAMAM-OH) and linear polymers had no or very minor effects only. The effect was seen at surprisingly low concentrations (7 µg mL^{-1} or below) and took place with no cytotoxicity and was demonstrated in two different systems; first, neuronal cell lines infected with misfolded prion protein and therefore continuously secreting PrP aggregates could be cured by several dendrimer types, in a dose- and time-dependent way, – 1.5 µg mL^{-1} treatment for 1 week removed all aggregated prion protein. Treatment of the cells with dendrimers caused the protease-resistant form of the prion protein to disappear and it did not reappear upon removal of the dendrimer. Secondly, dendrimers could render protease-resistant PrP aggregates in brain homogenates from scrapie RML (a specific, pathogenic prion strain) infected mice protease sensitive after treatment at pH<4 using the dendrimer at 60 μg mL^{-1}. PAMAM from and above G3, superfect (Qiagen), PPI G4 and average and high molecular weight PEI showed such effects.

The dendrimers were speculated to work similarly in both systems which led Supattapone and co-workers[111] to suggest that the action takes place in an acidic compartment of the cell. However, no direct evidence for this is presented and another explanation could be that the chronically infected cell lines are far more sensitive to the actions of the dendrimers than the preformed aggregates of brain homogenates simply because it is a living system and therefore even the minimally charged dendrimers existing at the neutral pH of the cell culture medium would have a discernible effect.

It was evident from testing a range of different pathogenic prion strains in brain homogenates from infected normal hamsters or mice or from infected transgenic mice that some strains were not sensitive to treatment with dendrimers, while others were. Interestingly, exposure to a denaturating agent like urea assisted in the dissolution of PrP aggregates, rendering some of the resistant strains susceptible to dendrimers. The pathogenic aggregated prion from mad cow disease (BSE) was much more susceptible to G4 PPI than scrapie prions (scrapie is a sheep prion disease not harmful to man).

Thus, the unfolding tendency or ease of unfolding differs between different prion strains and therefore dendrimers may be designed to target only certain prion strains whereby dendrimers become of potential use as a means to diagnose infections with different prion strains. This is of obvious importance when considering that certain prion infections are zoonotic (*i.e.* animal infections also infecting humans), while others are not.

The findings of Supattapone were reproduced and extended by Solassol and co-workers,[112] using cationic phosphorus-containing G4 and G5 dendrimers with tertiary amine surface groups (Figure 4.15), which were also found to work at non-cytotoxic concentrations. In addition to showing the prion curing effect in infected cell lines and the ability to solubilise prion aggregates in brain homogenates, this work furthermore shows that Prp aggregate formation could also be inhibited *in vivo* in the spleens of mice subjected to aggregated scrapie prion protein injected intraperitoneally and then injected with dendrimer every second day for 30 days. No adverse effects of administering these dendrimers to the mice were observed during this period of treatment. Also, using an *in vitro* infectivity assay it was shown that dendrimer-treated neuronal cell lines previously producing infectious prion proteins were not able to infect new cells. Finally, these dendrimers were found to work well at neutral pH obviously obviating the need for an acidic cell compartment.

The solubilising effect of dendrimers has been reproduced with prion protein derived, amyloid fibril forming peptides and it was shown that guanidino derivatised PPI dendrimers retained the fibril solubilising ability at neutral pH, as opposed to a non-derivatised (*i.e.* amino terminated) PPI dendrimer of the same size (G2), supporting that charged surface groups are a prerequisite for the protein solubilising effect of the dendrimer.[114]

Ottaviani and co-workers[113] studied the interactions of PAMAM dendrimers with model proteins and found the interactions to involve protonated surface amino groups and both charged protein surface groups and hydrophobic groups, and a

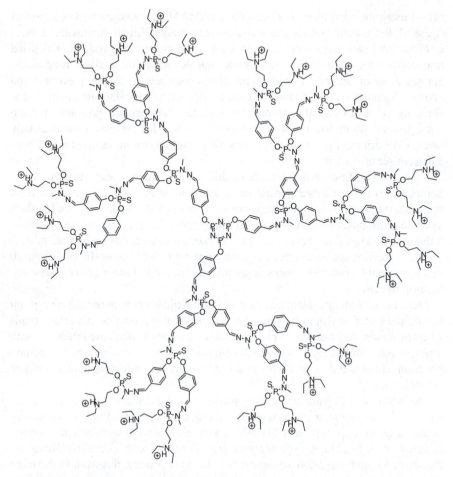

Figure 4.15 *A phosphorous-based dendrimer for the solubilisation of misfolded prion proteins developed by Solassol and co-workers[112]*

direct binding could be demonstrated by physical techniques (electron paramagnetic resonance spectroscopy).

Zhou and Ghosh[102] showed that a dendrimer with four copies of a leucine-zipper (*i.e.* amphipathic α-helix forming) peptide was able to form a complex with four complementary α-helical peptides resulting in a dendrimer displaying four coiled coil peptide dimers at the surface, also attesting to the ability of dendrimers to interact directly with peptide moieties at its surface.

Supattapone and co-workers[111] report that PPI binds directly to the protease-resistant core of misfolded PrP and a similar binding is reported by Solassol and co-workers.[112] Thus, a direct interaction between these dendrimers and the misfolded prion protein leading to a dissociation of PrP aggregates could be part of the mechanism of the aggregate solubilising effect of cationic dendrimers. It is however also feasible that at least part of the effect may be ascribed to a general solvent effect of the dendrimer *e.g.* by acting as a water structure perturbing solute (chaotrope), lowering the dielectric constant and the viscosity of water and disordering the regular water structure by reorganising water molecules at the dendrimer surface. As with other chaotropes, this would lead to hydrophobic interactions being disfavoured, which is highly destabilising for most protein tertiary structures. Classical examples of chaotrophic salts are $MgCl_2$, urea, guanidinium chloride, sodium thiocyanate and guanidinium thiocyanate at high concentrations and other chaotropes include polarity-decreasing, water miscible organic solvents like acetonitrile, propanol and methanol. Generally, chaotropes will solubilise proteins, which are useful, for example, in solubilising protein aggregates as are often encountered when expressing proteins in heterologous expression systems (inclusion body formation) and when extracting certain types of membrane proteins. Dendrimers, being typically compact, large polyionic substances have the physicochemical properties needed to make them potential chaotropes/protein denaturants. As noted above, Supattapone and co-workers[111] found the effects of urea and dendrimers to be additive.

The protein solubilising effect of certain dendrimers as established for prion protein dissolution could very well be a general property of large, compact polycationic dendrimers, but this awaits further testing with a wide range of dendrimers of different sizes and charge. Also other types of protein aggregates should be investigated, including amyloid protein and polypeptide aggregates and inclusion bodies from heterologously expressed recombinant proteins. This could open up completely new medical and biotechnological areas for applications of dendrimers. The principle of dendrimer-mediated protein aggregate solubilisation has seemingly not been applied to such important types of protein aggregates as *e.g.* Aβ-amyloid of Alzheimer's disease.

4.8 Summary

Dendrimer drugs are still in their infancy. However, many applications have been demonstrated, most of all attesting to the versatility of these compounds. Thus, groups with therapeutic potential can be presented on dendrimers in a multimeric fashion to increase their effect, dendrimers can be targeted to bind to specific types of cells only (*e.g.* neoplastic cells expressing specific receptor molecules) and dendrimers can be tailored to optimise the fit of a specific ligand with its receptor. Furthermore, dendrimers can be made to satisfy different size and solubility demands allowing their vascular survival *in vivo* and enabling them to reach different physiological compartments. Dendrimers may even be designed to decompose under specific conditions making the traceless delivery of a drug possible. Antimicrobial dendrimers generally rely either on the inhibition of attachment of the microbial agent or its toxin to the target host cells, or on selective lysis of the microbial cell membrane. Before dendrimers can be applied generally for *in vivo* uses as drugs, they have to be developed further to increase the efficiency by which they reach their target and to increase the specificity for the microbial target.

Dendrimer-based vaccines have shown great promise and have been almost solely based upon the peptide MAPs developed by Tam,[63,73] as these dendrimers have proven to be very versatile and able to accommodate a wide range of peptides and combinations of peptides as well as substances with adjuvant activity (lipidic structures). They have also consistently shown a surprising and very significant enhancement of peptide immunogenicity; in a number of cases, even MHC I-restricted cell-mediated immunity, important for the defence against viral infections or treatment of cancer cells has been achieved. There are indications that MAP constructs sometime bypass normal immune-restricting pathways and achieve efficient stimulation of the immune system without resorting to complicated structures or immunogen formulations. MAPs and similar dendrimeric constructs may therefore constitute the first real possibility of creating all-synthetic vaccines. Work needs to be done, however to reach the difficult goal of a simple, chemically defined immunogen that is at the same time efficient are in contradiction to the empirical finding that pure immunogens are poor immunogens. One aspect of this is to ensure that MAPs are really pure (defined by mass spectrometry); this is very often not reported. Another object will be to avoid the use of adjuvants or at least adjuvants like Freund's, which is not allowed for human use. Finally, it will be a challenge, when being able to produce 100% chemically defined immunogens to apply the knowledge on immune mechanisms in order to tailor-make the immunogen construct to obtain the desired type and magnitude of immune response.

The chance finding that certain dendrimers can act as protein denaturants needs to be explored further to elucidate if this is a general property of large, compact polycationic dendrimers or if this effect is restricted to interactions with aggregated prion proteins. Other types of protein aggregates should be investigated, including other amyloid proteins and inclusion bodies from heterologously expressed recombinant proteins. This could open up completely new medical and biotechnological areas for applications of dendrimers. The principle of dendrimer-mediated protein aggregate solubilisation has seemingly not been applied to such important types of protein aggregates as *e.g.* Aβ-amyloid of Alzheimer's disease.

References

1. T.K. Lindhorst, C. Kieburg and U. Krallmann-Wenzel, *Glycoconj. J.*, 1998, **15**, 605.
2. M. Kohn, J.M. Benito, C. Ortiz Mellet, T.K. Lindhorst and J.M.Garcia Fernandez, *Chembiochem*, 2004, **5**, 771.
3. M.L. Szalai, R.M. Kevwitch and D.V. McGrath, *J. Am. Chem. Soc.*, 2003, **125**, 15688.
4. F.M.H. de Groot, C. Albrecht, R. Koekkoek, P.H. Beusker and H.W. Scheeren, *Angew. Chem. Int. Ed.*, 2003, **42**, 4490.
5. R.J. Amir, N. Pessah, M. Shamis and D. Shabat, *Angew. Chem. Int. Ed.*, 2003, **42**, 4494.
6. D. Shabat, R.J. Amir, A. Gopin, N. Pessah and M. Shamis, *Chem. Eur. J.*, 2004, **10**, 2626.

7. N. Malik, R. Wiwattanapatapee, R. Klopsch, K. Lorenz, H. Frey, J.W. Weener, E.W. Meijer, W. Paulus and R. Duncan, *J. Control. Release*, 2000, **65**, 133.
8. H. Kobayashi and M.W. Brechbiel, *Mol. Imaging*, 2003, **2**, 1.
9. Z.-Y. Zhang and B.D. Smith, *Bioconjugate Chem.*, 2000, **11**, 805.
10. N. Langeland, L.J. Moore, H. Holmsen and L. Haarr, *J. Gen. Virol.*, 1988, **69**, 1137.
11. J.S. Aguilar, M. Rice and E.K. Wagner, *Virology*, 1999, **258**, 141.
12. Y. Gong, B. Matthews, D. Cheung, T. Tam, I. Gadawski, D. Leung, G. Holan, J. Raff and S. Sacks, *Antiviral Res.*, 2002, **55**, 319.
13. D.I. Bernstein, L.R. Stanberry, S. Sacks, N.K. Ayisi, Y.H. Gong, J. Ireland, R.J. Mumper, G. Holan, B. Matthews, T. McCarthy and N. Bourne, *Antimicrob. Agents Chemother.*, 2003, **47**, 3784.
14. J.D. Reuter, A. Myc, M.M. Hayes, Z. Gan, R. Roy, D. Qin, R. Yin, L.T. Piehler, R. Esfand, D.A. Tomalia and J.R. Baker Jr., *Bioconjugate Chem.*, 1999, **10**, 271.
15. F. Lasala, E. Arce, J.R. Otero, J. Rojo and R. Delgado, *Antimicrob. Agents Chemother.*, 2003, **47**, 3970.
16. M. Witvrouw, V. Fikkert, W. Pluymers, B. Matthews, K. Mardel, D. Schols, J. Raff, Z. Debyser, E. De Clercq, G. Holan and C. Pannecouque, *Mol. Pharmacol.*, 2000, **58**, 1100.
17. N. Bourne, L.R. Stanberry, E.R. Kern, G. Holan, B. Matthews and D.I. Bernstein, *Antimicrob. Agents Chemother.*, 2000, **44**, 2471.
18. Y.H. Jiang, P. Emau, J.S. Cairns, L. Flanary, W.R. Morton, T.D. McCarthy and C.C. Tsai, *AIDS Res. Hum. Retroviruses*, 2005, **21**, 207.
19. M. Witvrouw, H. Weigold, C. Pannecouque, D. Schols, E. DeClercq and G. Holan, *J. Med. Chem.*, 2000, **43**, 778.
20. A. Gazumyan, B. Mitsner and G.A. Ellestad, *Curr. Pharm. Des.*, 2000, **6**, 525.
21. V. Razinkov, A. Gazumyan, A. Nikitenko, G. Ellestad and G. Krishnamurthy, *Chem. Biol.*, 2001, **8**, 645.
22. R.D. Kensinger, B.J. Catalone, F.C. Krebs, B. Wigdahl and C.L. Schengrund, *Antimicrob. Agents Chemother.*, 2004, **48**, 1614.
23. J.J. Landers, Z. Cao, I. Lee, L.T. Piehler, P.P. Myc, A. Myc, T. Hamouda, A.T. Galecki and J.R. Baker Jr., *J. Infect. Dis.*, 2002, **186**, 1222.
24. J. Rojo and R. Delgado, *J. Antimicrob. Chemother.*, 2004, **54**, 579.
25. H. Feldmann, S. Jones, H.D. Klenk and H.J. Schnittler, *Nat. Rev. Immunol.*, 2003, **33**, 677.
26. R.W. Pinner, S.M. Teutsch, L. Simonsen, L.A. Klug, J.M. Graber, M.J. Clarke and R.L. Berkelman, *J. Am. Med. Assoc.*, 1996, **275**, 189.
27. R. Medzhitov and C. Janeway Jr., *Trends Microbiol.*, 2000, **8**, 452.
28. K.V. Reddy, R.D. Yedery and C. Aranha, *Int. J. Antimicrob. Agents*, 2004, **24**, 536.
29. Y. Shai, *Biochim. Biophys. Acta*, 1999, **1462**, 55.
30. A. Mecke, S. Uppuluri, T.M. Sassanella, D.K. Lee, A. Ramamoorthy, J.R. Baker Jr., B.G. Orr and M.M. Banaszak Holl, *Chem. Phys. Lipids*, 2004, **132**, 3.
31. S. Hong, A.U. Bielinska, A. Mecke, B. Keszler, J.L. Beals, X. Shi, L. Balogh, B.G. Orr, J.R. Baker Jr. and M.M. Banaszak Holl, *Bioconjugate Chem.*, 2004, **15**, 774.

32. C.Z. Chen and S.L. Cooper, *Biomaterials*, 2002, **23**, 3359.
33. V. Percec, A.E. Dulcey, V.S.K. Balagurusamy, Y. Miura, J. Smidrkal, M. Peterca, S. Nummelin, U. Edlund, S. Hudson, P.A. Heiney, H. Duan, S.N. Magonov and S.A. Vinogradov, *Nature*, 2004, **430**, 764.
34. C.Z. Chen, N.C. Beck-Tan, P. Dhurjati, T.K. van Dyk, R.A. LaRossa and S.L. Cooper, *Biomacromolecules*, 2000, **1**, 473.
35. K.A. Karlsson, *Adv. Exp. Med. Biol.*, 2001, **491**, 431.
36. C.L. Schengrund, *Biochem. Pharmacol.*, 2003, **65**, 699.
37. J.P. Tam, Y.A. Lu and J.L.Yang, *Eur. J. Biochem.*, 2002, **269**, 923.
38. N. Nagahori, R.T. Lee, S. Nishimura, D. Page, R. Roy and Y.C. Lee, *Chembiochem*, 2002, **3**, 836.
39. I. Vrasidas, N.J. de Mol, R.M. Liskamp and R.J. Pieters, *Eur. J. Org. Chem.*, 2001, **66**, 4685.
40. E.A. Merritt, Z. Zhang, J.C. Pickens, M. Ahn, W.G. Hol and E. Fan, *J. Am. Chem. Soc.*, 2002, **124**, 8818.
41. K. Nishikawa, K. Matsuoka, M. Watanabe, K. Igai, K. Hino, K. Hatano, A. Yamada, N. Abe, D. Terunuma, H. Kuzuhara and Y. Natori, *J. Infect. Dis.*, 2005, **191**, 2097.
42. T.P. Devasagayam and J.P. Kamat, *Indian J. Exp. Biol.*, 2002, **40**, 680.
43. S.H. Battah, C.E. Chee, H. Nakanishi, S. Gerscher, A.J. MacRobert and C. Edwards, *Bioconjugate Chem.*, 2001, **12**, 980.
44. N. Nishiyama, H.R. Stapert, G.D. Zhang, D. Takasu, D.L. Jiang, T. Nagano, T. Aida and K. Kataoka, *Bioconjugate Chem.*, 2003, **14**, 58.
45. H. Kobayashi and M.W. Brechbiel, *Mol. Imaging*, 2003, **2**, 1.
46. K. Kono, M. Liu and J.M. Frechet, *Bioconjugate Chem.*, 1999, **10**, 1115.
47. M.G. Baek and R. Roy, *Bioorg. Med. Chem.*, 2002, **10**, 11.
48. R. Roy, M.G. Baek and O. Rittenhouse, *J. Am. Chem. Soc.*, 2001, **123**, 1809.
49. R. Roy and M.G. Baek, *J. Biotechnol.*, 2002, **90**, 291.
50. V. Berezin and E. Bock, *J. Mol. Neurosci.*, 2004, **22**, 33.
51. J.L. Neiiendam, L.B. Kohler, C. Christensen, S. Li, M.V. Pedersen, D.K. Ditlevsen, M.K. Kornum, V.V. Kiselyov, V. Berezin and E. Bock, *J. Neurochem.*, 2004, **91**, 920.
52. K. Cambon, S.M. Hansen, C. Venero, A.I. Herrero, G. Skibo, V. Berezin, E. Bock and C. Sandi, *J. Neurosci.*, 2004, **24**, 4197.
53. J. Wu, J. Zhou, F. Qu, P. Bao, Y. Zhang and L. Peng, *Chem. Commun.*, 2005, **3**, 313.
54. H. Zhao, J. Li, F. Xi and L. Jiang, *FEBS Lett.*, 2004, **563**, 241.
55. N. Nitin, P.J. Santangelo, G. Kim, S. Nie and G. Bao, *Nucl. Acids Res.*, 2004, **32**, e58.
56. C.A. Janeway, P. Travers, M. Walport and M. Shlomchik in *Immunobiology*, 5th edn, Garland, New York, 2001.
57. S.F. Slovin, S.J. Keding and G. Ragupathi, *Immunol. Cell Biol.*, 2005, **83**, 418–428.
58. I. Dalum, D.M. Butler, M.R. Jensen, P. Hindersson, L. Steinaa, A.M. Waterston, S.N. Grell, M. Feldmann, H.I. Elsner and S. Mouritsen, *Nature Biotechnol.*, 1999, **17**, 666.

59. L.J. Patterson, F. Robey, A. Muck, K. van Remoortere, K. Aldrich, E. Richardson, W.G. Alvord, P.D. Markham, M. Cranage and M. Robert-Guroff, *AIDS Res. Hum. Tetroviruses*, 2001, **17**, 837.

60. D.C. Jackson, N. O'Brien-Simpson, N.J. Ede and L.E. Brown, *Vaccine*, 1997, **15**, 1697.

61. M.H. van Regenmortel, J.P. Briand, S. Muller and S. Plaue, in *Synthetic Polypeptides as Antigens*, Elsevier, Amsterdam, 1988.

62. C. Plank, K. Mechtler, F.C. Szoka and E. Wagner, *Hum. Gene Ther.*, 1996, **7**, 1437.

63. J.P. Tam, *Proc. Natl. Acad. Sci. USA*, 1988, **85**, 5409.

64. K. Sadler and J.P. Tam, *Rev. Mol. Biotechnol.*, 2002, **90**, 195.

65. J.P. Defoort, B. Nardelli, W. Huang and J.P. Tam, *Int. J. Peptide Protein Res.*, 1992, **40**, 214.

66. D.N. Posnett, H. McGrath and J.P. Tam, *J. Biol. Chem.*, 1988, **263**, 1719.

67. C.A. Moreno, R. Rodriguez, G.A. Oliveira, V. Ferreira, R.S. Nussenzweig, Z.R.M. Castro, J.M. Calvo-Calle and E. Nardin, *Vaccine*, 2000, **18**, 89.

68. E.H. Nardin, J.M. Calvo-Calle, G.A. Oliveira, R.S. Nussenzweig, M. Schneider, J.M. Tiercy, L. Loutan, D. Hochstrasser and K. Rose, *J. Immunol.*, 2001, **166**, 481.

69. E.H. Nardin, G.A. Oliveira, J.M. Calvo-Calle, Z.R. Castro, R.S. Nussenzweig, B. Schmeckpeper, B.F. Hall, C. Diggs, S. Bodison and R. Edelman, *J. Infect. Dis.* 2000, **182**, 1486.

70. J.P. Briand, C. Barin, M.H. van Regenmortel and S. Muller, *J. Immunol. Meth.*, 1992, **156**, 255.

71. I. Haro and M.J. Gomara, *Curr. Protein Peptide Sci.*, 2004, **5**, 425.

72. P. Veprek and J. Jezek, *J. Peptide Sci.*, 1999, **5**, 203.

73. J.P. Tam, *J. Immunol. Meth.*, 1996, **196**, 17.

74. E.H. Nardin, G.A. Oliveira, J.M. Calvo-Calle and R.S. Nussenzweig, *Adv. Immunol.*, 1995, **60**, 105.

75. G. Esposito, F. Fogolari, P. Viglino, S. Cattarinussi, M.T. De Magistris, L. Chiappinelli and A. Pessi, *Eur. J. Biochem.*, 1993, **217**, 171.

76. R. Wang, Y. Charoenvit, G. Corradin, R. Porrozzi, R.L.Hunter, G. Glenn, C.R. Alving, P. Church and S.L. Hoffman, *J. Immunol.*, 1995, **154**, 2784.

77. I. Toth, M. Danton, N. Flinn and W.A. Gibbons, *Tetrahedron Lett.*, 1993, **34**, 3925.

78. C. Olive, M. Batzloff, A. Horvath, T. Clair, P. Yarwood, I. Toth and M.F. Good, *Infect. Immun.*, 2003, **71**, 2373.

79. K. Deres, H. Schild, K.H. Wiesmüller, G. Jung and H.G. Rammensee, *Nature*, 1989, **342**, 561.

80. J.S. Cavenaugh, H.-K. Wang, C. Hansen, R.S. Smith and J.N. Herron, *Pharm. Res.*, 2003, **4**, 591.

81. W. Olszewska, O.E. Obeid and M.W. Steward, *Virology*, 2000, **272**, 98.

82. S. Ota, T. Ono, A. Morita, A. Uenaka, M. Harada and E. Nakayama, *Cancer Res.*, 2002, **62**, 1471.

83. J.P. Tam and J.C. Spetzler, *Biomed. Peptide Proteins Nucl. Acids*, 1995, **1**, 123.

84. W.C. Chan and P.D. White, *Fmoc Solid Phase Peptide Synthesis*, Oxford Univeristy Press, Oxford, 2000.

85. J.C. Spetzler and J.P. Tam, *Peptide Res.*, 1996, **9**, 190.
86. E. de Oliveira, J. Villen, E. Giralt and D. Andreu, *Bioconjugate Chem.*, 2003, **14**, 144.
87. M.A. Rigby, N. Mackay, G. Reid, R. Osborne, J.C. Neil and O. Jarrett, *Vaccine*, 1996, **14**, 1095.
88. P.M.H. Heegaard, Y.S. Bøg, P.H. Jakobsen and A. Holm, in *Innovation and Perspectives in Solid Phase Synthesis*, R. Epton (ed), Mayflower Scientific Ltd, Birmingham, 1998, 143.
89. C. Olive, K. Hsien, A. Horvath, T. Clair, P. Yarwood, I. Toth, M.F. Good, *Vaccine*, 2005, **23**, 2298–2303.
90. K. White, P. Kearns, I. Toth and S. Hook, *Immunol. Cell Biol.*, 2004, **82**, 517.
91. B. Nardelli, P.B. Haser and J.P. Tam, *Vaccine*, 1996, **14**, 1335.
92. E. Iglesias, J.C. Aguilar, L.J. Cruz and O. Reyes, *Mol. Immunol.*, 2005, **42**, 99.
93. L.J. Cruz, E. Iglesias, J.C. Aguilar, A. Cabrales, O. Reyes and D. Andreu, *Bioconjugate Chem.*, 2004, **15**, 112.
94. M.G. Baek and R. Roy, *Bioorg. Med. Chem.*, 2002, **10**, 11.
95. S. Bay, R. Lo-Man, E. Osinaga, H. Nakada, C. Leclerc and D. Cantacuzene, *J. Peptide Res.*, 1997, **49**, 620.
96. R. Lo-Man, S. Vichier-Guerre, S. Bay, E. Deriaud, D. Cantacuzene and C. Leclerc, *J. Immunol.*, 2001, **166**, 2849.
97. T. Toyokuni, S. Hakomori and A.K. Singhal, *Bioorg. Med. Chem.*, 1994, **2**, 1119.
98. L. Vannucci, A. Fiseriva, K. Sadalapure, T.K. Lindhorst, M. Kuldova, P. Rossmann, O. Horvath, V. Kren, P. Krist, K. Bezouska, M. Luptovcova, F. Mosca and M. Popisil, *Int. J. Oncol.*, 2003, **23**, 285.
99. G. Mihov, D. Grebel-Koehler, A. Lubbert, G.W. van der Meulen, A. Herrmann, H.A. Klok and K. Mullen, *Bioconjugate Chem.*, 2005, **16**, 283–293.
100. V. Tsikaris, C. Sakarellos, M. Sakarellos-Daitsiotis. M.T. Chung, M. Marraud, G. Konidou, A. Tzinia and K.P. Soteriadou, *Peptide Res.*, 1996, **9**, 240.
101. G. Tuchscherer, B. Dömer, U. Sila, B. Kamber and M. Mutter, *Tetrahedron*, 1993, **49**, 3559–3575.
102. M. Zhou and I. Ghosh, *Org. Lett.*, 2004, **6**, 3561.
103. R.P. McGeary, I. Jablonkai and I. Toth, *Tetrahedron*, 2001, **57**, 8733.
104. P. Rajananthanan, G.S. Attard, N.A. Sheikh and W.J. Morrow, *Vaccine*, 1999, **17**, 715.
105. B. Morein, M. Villacres-Eriksson and K. Lovgren-Bengtsson, *Dev. Biol. Stand.*, 1998, **92**, 33.
106. D.C. Wright, *Columbia Patent US005795582A*, 1998, p. 7.
107. E.J. Baird, D. Holowka, G.W. Coates and B. Baird, *Biochemistry*, 2003, **11**, 12739.
108. S. Shaunak, S. Thomas, E. Gianasi, A. Godwin, E. Jones, I. Teo, K. Mireskandari, P. Lubert, R. Duncan, S. Patterson, P. Khaw and S. Brocchini, *Nature Biotech.*, 2005, **22**, 977.
109. A.M. Krieg, *Springer Semin. Immunopathol.*, 2000, **22**, 97.
110. S. Supattapone, H.-O.B. Nguyen, F.E. Cohen, S.B. Prusiner and M. Scott, *Proc. Natl. Acad. Sci., USA*, 1999, **96**, 14529.

111. S. Supattapone, H. Wile, L. Uyechi, J. Safar, P. Tremblay, F.C. Szoka, F.E. Cohen, S.B. Prusiner and M.R. Scott, *J. Virol.*, 2001, **75**, 3453.
112. J. Solassol, C. Crozet, V. Perrier, J. Leclaire, F. Béranger, A.-M. Caminade, B. Meunier, D. Dormont, J.-P. Majoral and S. Lehman, *J. Gen. Virol.*, 2004, **85**, 1791.
113. M.F. Ottaviani, S. Jockusch, N.J. Turro, D.A. Tomalia and A. Barbon, *Langmuir*, 2004, **20**, 10238.
114. P.M.H. Heegaard, H.G. Pedersen, J. Flink and U. Boas, *FEBS Lett.*, 2004, **577**, 127.

CHAPTER 5

Dendrimers in Diagnostics

5.1   Contrast Agents based on Dendrimers

Contrast agents have a long history in medicine as a tool for achieving enhanced resolution in imaging of different internal structures in the organism. The classical example is the use of barium sulfate as a contrast agent for X-ray investigation of the digestive system. The methods for imaging living biological structures have undergone a tremendous development during the last 30 years, starting from the classical use of X-ray pictures to the present methods such as computer-aided tomography (CT), scintigraphy and magnetic resonance imaging (MRI). CT and MRI gives 3-D pictures of the object investigated, while scintigraphy, which rely on emission from a radioactive compound, may give information on metabolic, biochemical and functional activity.

   All X-ray techniques rely on the attenuation of X-rays by the tissue, as a consequence, tissues containing heavier elements such as calcium in bones have a higher natural contrast than soft tissues. The consequence of this is that the contrast agents for X-ray-based techniques must incorporate heavier elements to achieve a better contrast. An element with a reasonable balance between toxicity and contrast is iodine, therefore, the majority of contrast agents for CT contains iodine.

   An MRI is based on the mapping of proton density in the tissue by the use of nuclear magnetic resonance (NMR). The subject is placed in a magnetic field, which induces an orientation of the nucleic spins of ^1H present in the tissue. The orientation of the ^1H-spins follows a Boltzman distribution, where a change in the orientation towards the magnetic field is further stimulated by an applied radio pulse. This change in distribution of the ^1H-spins is detected and constitutes the MRI signal. The relaxation (or reorientation) of the ^1H-nuclei determines the contrast in the pictures, and the contrast agents for MRI act by shortening the relaxation times (T1 and T2) of the ^1H nuclei. An NMR experiment takes place in 3-D with an orientation induced by the external magnetic field and T1 describes relaxation taking place in the field direction (the z-direction) and is also commonly called the spin lattice or longitudinal relaxation constant. T2 describes the relaxation processes taking place in the X–Y plane and is also called the spin–spin or transversal relaxation constant. The contrast agents can be classified according to their mechanism of action: Gd^{3+} and Mn^{3+} affects T1, while magnetic iron oxides (magnetites) and Dy^{3+} affects T2. The problem encountered with contrast agents for MRI is toxicity of metals such as Gd^{3+} and Dy^{3+} as well as biological activity (and risk of poisoning) of Mn in various oxidation states, and this has been addressed by using chelating ligands for binding of the metal ions.

Dendrimers offer several advantages, when considering contrast agents: the dendritic architecture is highly suitable for incorporation of many contrast-giving atoms/ions while keeping the solubility. Dendrimers are accessible in sizes that allow imaging of vascular structures, which are not easily imaged by low-molecular weight contrast agents due to fast diffusion (extravasation) of the small molecule agent into the surrounding tissue. Dendrimer-based contrast agents can also be made target specific by using dendrimers with a tag capable of binding to a specific type of tissue.

An important aspect is the amount of contrast agent necessary for getting the desired resolution. CT requires from 100 to 1000 mg kg^{-1}, MRI from 0.1 to 0.001 mg kg^{-1} and scintigraphy from 10^{-5} to 10^{-8} mg kg^{-1}.[1–5]

5.1.1 Dendrimer-based Contrast Agents for CT

There are only few reports in the literature on dendrimer-based contrast agents for CT. Iodine-containing dendrimers have been described by a couple of groups,[3,6] but the somewhat definitive study on the synthesis and characterisation of dendrimer-based X-ray contrast agents must be ascribed to the group at Schering AG in Berlin,[3] which used 2,4,6-triiodophenyl-group as a contrast yielding unit attached to the surface of PAMAM-, PPI- or polylysine-dendrimers (Figure 5.1).

The thermal stability of the dendrimers in solution was investigated, since sterilisation is essential for all parenteral drugs. The PAMAM dendrimers were not stable in this treatment, whereas both PPI- and polylysine-dendrimers showed stability. The toxicity and clearance in mice were also investigated, and it was only the polylysine dendrimer that showed low toxicity and low retention. However, the conclusion was that the polylysine dendrimer-based contrast agents are not likely to compete on commercial terms with the commercial Gd-based MRI contrast agent Gadomer-17 (also developed by Schering AG), since the amount of material necessary for CT (gram amounts) is much larger than what is necessary for an MRI and the spatial resolution is often satisfactory in an MRI.

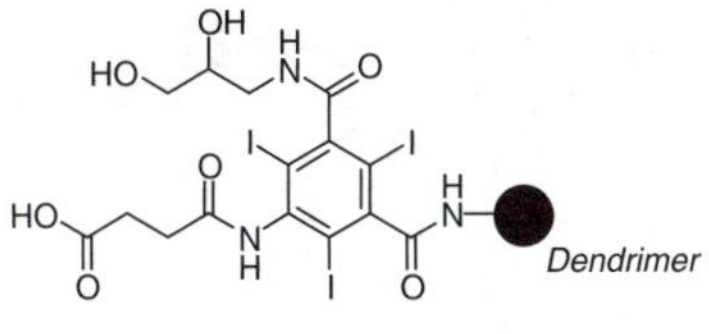

a 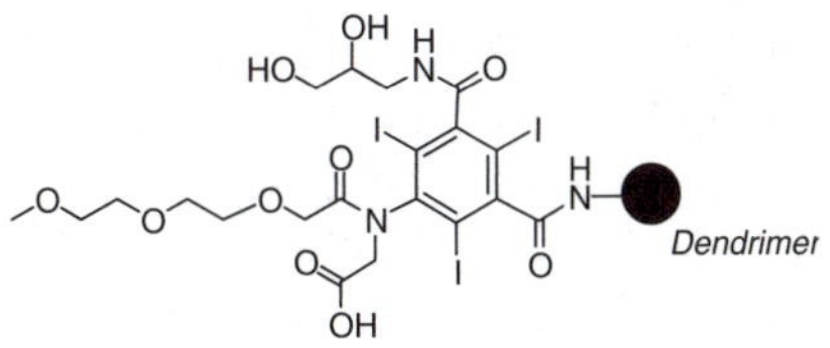   b

Figure 5.1 *Two examples of iodinated organic compounds used as dendrimer conjugates by Krause and co-workers[3] as CT-contrast agents*

There have also been reports on organotin-[7] and organobismuth- based dendrimers[8] as potential contrast agents, however, it is more speculative if these reagents will find clinical use due to the known toxicity of organotin compounds[9] and the possible toxicity of organobismuth compounds.[10]

5.1.2   Dendrimer-based Contrast Agents for MRI

These reagents may be divided into two classes of materials: those influencing T1 and those influencing T2. The majority of contrast agents available at present belong to the class of T1 materials, and are essentially based on metal ions such as Gd^{3+} and Mn^{2+} or Mn^{3+} bound to chelating ligands with a labile site for coordination of water molecules. The problem is to achieve the best balance between stability of the complex and the rate of water exchange. The most commonly used ligands are from the EDTA family (Figure 5.2).

The first dendrimer-based contrast agents for an MRI as described by Wiener and co-workers,[11,12] were PAMAM dendrimers with covalently bound DTPA ligands (Figure 5.2c) for binding Gd^{3+}. They demonstrated the usefulness of these dendrimers for MR angiography (visualisation of vascular structures). Blood circulation half-lives ranged from 40 to 200 min depending on the molecular weight of the dendrimer used. Another type of Gd^{3+}-binding dendrimer based on the ligand DO3A (Figure 5.2b) bound to a PAMAM dendrimer has been investigated, and showed that the minimal dose necessary for visualisation of vascular structures in rabbits was 0.02 mmol kg^{-1} animal corresponding to 1.16 g kg^{-1}.[13]

Kobayashi and co-workers[14,15] investigated the pharmacokinetics of dendrimer-based MRI contrast agents using PPI and PAMAM dendrimers (G2–G4) carrying DTPA groups (Figure 5.2a) as surface groups in order to bind the gadolinium ion used for contrast imaging. Although having almost identical molecular weights, PPI dendrimer-based agents were found to be more rapidly excreted from the body than the agents based on PAMAM dendrimers having same number of surface groups and, not surprisingly smaller dendrimers were more rapidly excreted than large dendrimers; for contrast reagent use, G3-PAMAM and G3-PPI with 16 end groups were optimal. A further interesting feature is that the contrast agents based on PPI dendrimers have

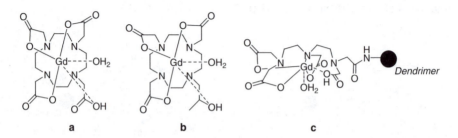

      a   b   c

Figure 5.2 *The three most commonly used chelating ligands used for metal binding in MRI contrast agents shown with Gd^{3+} as the metal center and the exchangeable water molecule coordinated to the metal center. The ligands are: a: DOTA; b: DO3A; c: DTPA (conjugated to a dendrimer)*

higher relaxivity than the corresponding PAMAM dendrimers, and thus are more efficient.[16] This is ascribed to the differences in molecular shape between PAMAM and PPI dendrimers. The PPI dendrimers are more ellipsoid shaped than PAMAM dendrimers leading to a higher T1 relaxivity.

As with any contrast agent, the agent should be present in circulation long enough to allow visualisation, for example, of blood vessels but not so long as to present a potential hazard to the body. In addition to renal excretion, extravasculation also contributes to lowering the retention time in the circulation. Some dendrimers also accumulate in the kidneys before excretion, which may be useful for visualising renal structures.[14]

Dendrimers are due to their multivalent surface suitable as platforms for conjugates both carrying a contrast agent as well as a tag targeting a specific type of tissue, for detection of a specific type of tissue in areas, where the contrast otherwise would be poor. This has been demonstrated by Brechbiel and co-workers,[17,18] who utilised the overexpression of folate receptors on cancer cells for tagging with dendrimers carrying folic acid as well as a contrast agent.

The first commercial dendrimer-based contrast agent is Gadomer-17, which has been developed by Schering AG. It is based on a polylysine core substituted with DTPA groups at the surface and is presumably going to be marketed as a contrast agent for MR angiography.

5.1.3 Dendrimer-based Contrast Agents for Scintigraphy

Scintigraphy is a technique based on the detection of γ-rays irradiating from a radioactive element. The method is suitable for measuring metabolic or biochemical activities, but rely on access to the suitable carriers for the radioactive element used. One of the few examples of dendritic carriers in this area is the dendritic porphyrins (Figure 5.3) described by Mukhtar and co-workers.[19] Porphyrins are known to concentrate in tumour tissue, where they may serve as photosensitizers for photodynamic therapy (see Chapter 4). By tagging them with ^{99m}Tc they can be utilised for the localisation of tumours. In their study, they showed that the two dendrimers were concentrated in C_6-glioma tumours in the brain of rats.

5.2 Fluorescence Enhanced by Dendrimers

Fluorescence is a popular method in bioanalytical chemistry due to high sensitivity. However, in some cases it would be desirable to have an even higher sensitivity.

Two strategies for fluorescence-based detection of an analyte commonly used are: the reaction between an analyte and a fluorescent tag under conditions where an excess of the fluorescent tag can be removed. The remaining fluorescence correlates with the amount of analyte present. In this case, signal amplification can be obtained by increasing the number of fluorophores bound to the tag giving a larger signal to detect. The other strategy is to have a substrate for a specific reaction, where the substrate has a fluorescent group as well as a fluorescence quencher positioned at such a distance that the substrate becomes non-fluorescent. This phenomena is called fluorescence resonance energy transfer (FRET), and can also take place between multiple copies of

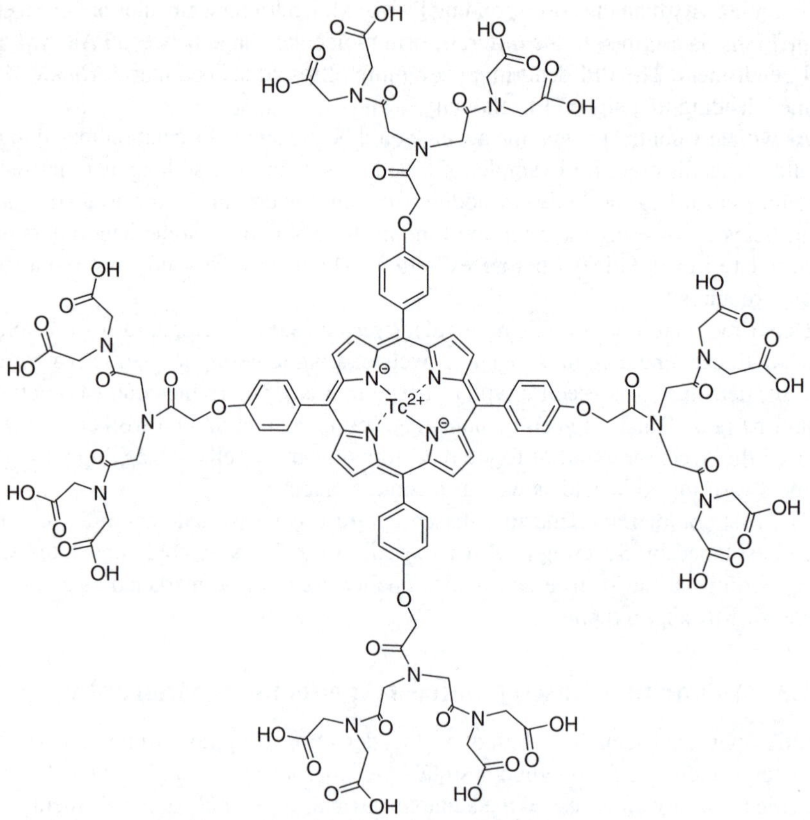

Figure 5.3 *A G2-porphyrin-core dendrimer used by Mukhtar and co-workers[19] as carrier for ^{99m}Tc for tumour localisation by scintigraphy*

the same fluorophore due to the energy loss associated with the transfer process. When the reaction in question takes place, the two parts of the molecule are separated, and fluorescence from the fluorophore can be observed. This approach could also be called the dendritic beacon.

The use of dendrimers for enhancing sensitivity in a human-herpes assay based on DNA microarrays has been reported. The assay is based on polymerase chain reaction (PCR) amplification of the viral DNA obtained from the patient, but uses fluorescent dendrimer-labelled primers carrying up to three fluorophores per oligonucleotide, followed by microarray analysis (*vide infra*).[20]

The dendritic beacon has been utilised by the groups of McIntyre[21] and Bradley[22] for assaying proteases. McIntyre's group were interested in an *in vivo* detection of matrix metalloprotease-7, which is associated with benign intestinal tumours. The assay was based on fluorescein (FITC)-labelled PAMAM dendrimers conjugated to a tetramethylrhodamine (TMR)-labelled peptide. When the two different fluorophores are present in the same molecule, only fluorescence from TMR can be observed. When the peptide is degraded by the enzyme, this splits off the part of the dendrimer carrying the FITC label from the peptide carrying the TMR label, and the characteristic fluorescence of FITC is observed. The assay was shown to work in a mouse model, but the sensitivity compared to a non-dendrimer system was not addressed.

Bradley and co-workers[22] developed a system for screening of substrates for proteases based on FRET between fluorophores and a quencher in peptide libraries (*vide infra*).

5.3 Dendrimers in Bioassays

Bioassays are used for identification, characterisation and/or quantitation of cells and biomolecules, including proteins, nucleic acids and carbohydrates and also include methods for determination of the biological activity, for example, of enzymes, cell-stimulating substances and toxins.

Dendrimers may be useful and versatile tools for increasing the sensitivity of bioassays in two major ways; first, dendrimers can be used to enhance signal generation, taking advantage of the high number of surface groups available for coupling of signal generating moieties and of the possibility to heterofuntionalise the dendrimer. This allows coupling both to multiple signal generating moieties and to the biospecific detection molecule (antibodies, oligonucleotides, receptor ligands, *etc.*) that determines the specificity and forms the core of the bioassay. Biospecific interactions may also be included as an intrinsic part of the signal generation step. Second, dendrimers can be used to enhance the covalent-binding capacities of solid phases used for heterogeneous assays like enzyme-linked immunosorbent assays (ELISA, typically polystyrene, see Figure 5.4), microarrays (typically glass) and plasmon surface resonance spectroscopy (typically gold). The analytical assays traditionally utilise non-covalent binding of analytes or detection molecules to the surface, however, covalency may increase the stability, reusability and selectivity of the assay. To increase the assay sensitivity it is, however, often necessary to increase the covalent-binding capacity of the surface, and dendrimers having multivalent surfaces are well suited for this purpose.

5.3.1 Dendrimer-Enhanced Signal Generation

DNA-based dendrimers are examples of highly amplified signal generating entities for general labelling of DNA molecules and other molecules equipped with a complementary DNA strand. The all-DNA dendrimer type, first described by Nilsen and co-workers[23](see Chapter 1) was used in combination with a signal generating oligonucleotide construct called a "signal amplification cassette" (SAC) and a polymerase/exonuclease method called "nucleotide extension and excision coupled signal amplification" (NEESA), (Figure 5.5). When coupled with luciferase-based detection of inorganic pyrophosphate arising from the NEESA reaction, high signal intensities are achieved with a single dendrimer molecule.[24] The 3' ends for the SACs and the polymerase extension reaction are provided by the DNA dendrimer itself and are present in high numbers on the surface of the dendrimer (as described in Chapter 1). Given the right composition of the NEESA, SACs could be detected down to the atto-mole level (10^{-18} mol). However, coupling SACs to 4,6 or 8-layer DNA dendrimers,

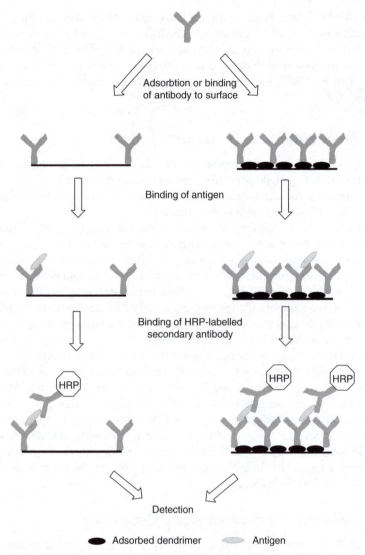

Figure 5.4 *A bioassay type employing a solid phase: ELISA with an antigen binding to a primary, specific solid phase-adsorbed antibody (the "catching" antibody) and detection by a second antibody (the "detection" antibody), labelled with a colour-generating enzyme*

potentially containing 162, 1457 and 13,122 potential surface groups, respectively, dramatically enhanced the sensitivity expressed as moles of $SAC_{multiple}$ dendrimer detectable by the method; this reached 5 zeptomoles (10^{-21} mol) for the 8-layer DNA dendrimer.[24] This immense sensitivity is thus achieved by combining catalytic amplification with stoichiometric amplification and is comparable to the amplification of nucleic acids achievable with established methods like PCR.

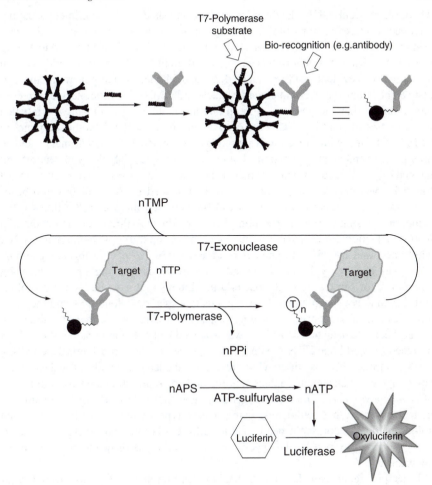

Figure 5.5 *DNA dendrimer[23] formed by specific annealing of complementary oligonucleotides and subsequent chemical crosslinking, exhibiting on its surface a number of free 5' and 3' oligonucleotide "arms" of known sequences onto which can be annealed oligonucleotides carrying molecules of interest; 3'-terminating arms can be used to attach a SAC label combined with biorecognition moieties of different kinds annealed to the 5'-"arms" of the dendrimer.[24] SACs can be used for polymerase catalysed extension/regeneration reactions, as they contain free 3'end and an over-hang of nucleotides. The isothermic NEESA reaction utilises the combined poly-merase and the 3'–5' exonuclease activity of certain DNA polymerases, generating pyrophosphate (PPi)*

Importantly, SAC-coupled DNA dendrimers also have a high number of surface-localised 5'-terminated DNA strands available for coupling to specific detection molecules for application in antibody based or other assays and thus can be used as general labelling reagents for high-sensitivity applications.

The Nilsen[23] DNA dendrimer has found widespread use for the enhancement of fluorescence- and enzyme-based signals, especially in microarray applications where fluorescent DNA dendrimers can be annealed to PCR primers carrying oligonucleotide extensions binding to the surface nucleotides present on the dendrimer, which allow the straightforward and general labelling of PCR products (see below). As these DNA dendrimers carry two different "free" oligonucleotide tags on the surface it is evident that they can be decorated with any signal-generating moiety that can be coupled to a complementary nucleotide. Also, any biomolecule or recognition moiety that can be coupled to an oligonucleotide tag can readily be labelled with this type of dendrimer. This has been demonstrated with antibiotin antibodies in ELISA (labelled with horseradish peroxidase decorated DNA dendrimer)[25] and in protein array and bead applications, both using dendrimers labelled with fluorescent molecules.[25,26] Sensitivity enhancements from 50- to 500- fold have been reported for such constructs. Using single-stranded biotinylated DNA amplicons in a bead-based assay, the signal-to-noise ratio can be increased more than 8.5 times when comparing a single-step streptavidin-phycoerythrin (SA-PE, fluorophore binding biotin) detection with detection mediated by a mixture of biotinylated DNA dendrimers and SA-PE.[26] In another bead assay, *Listeria monocytogenes* genomic DNA was caught by DNA probes attached to beads, making up a suspension microarray for sub-typing of the bacterium.[27] Interestingly, the beads can be colour coded such that a specific probe corresponds to a specific colour, allowing the rapid identification of sequences of binding beads in an extensive suspended library of beads. Adequate detection sensitivity was achieved without amplification of the DNA by reacting bead-bound DNA with another probe attached to a biotin labelled G4 DNA dendrimer then visualised by fluorescently labelled streptavidin. The dendrimer-based signal enhancement enables this method for rapid and extensive sub-typing of *L. monocytogenes*. The method has also been shown to work for the identification of several other pathogens, using a new type of dendrimer-like DNA and colour-encoded fluorescent beads having different colours and forming a decodable "barcode" formed by the specific number and types of beads included in the dendrimer.[28]

Detection signals can also be enhanced by dendrigraft-type (comb-like) polymers in which a primary nucleotide functions to hybridise to the target while the secondary nucleotides are synthesised as the "teeth" on the comb, inserted at certain branching points in the nucleotide also containing the primary oligonucleotide. By additional enzymatic ligations, it was possible to create multimers of these molecules that could be used for signal amplification by annealing the secondary nucleotides with alkaline phosphatase-labelled oligonucleotides for very sensitive detection of pathogen (viral) nucleic acid.[29]

An interesting application of dendrimers for the labelling of detection molecules like antibodies is metal-carbonyl-dendrimer immunoconjugates, which can be detected by the highly sensitive technique Fourier transform infrared (FTIR) spectroscopy (Figure 5.6).[30] In a model study, G5 PAMAM was coupled with the iron–carbonyl complex Fp-maleimide and then coupled to an antibody by reductive amination of antibody carbonyl groups (obtained by mild oxidation of the carbohydrate moieties of the antibody molecules) by the remaining dendrimer amino groups. Such antibodies were tested for detection of specific antigens adsorbed to nitrocellulose membranes and a good sensitivity and low background was obtained.

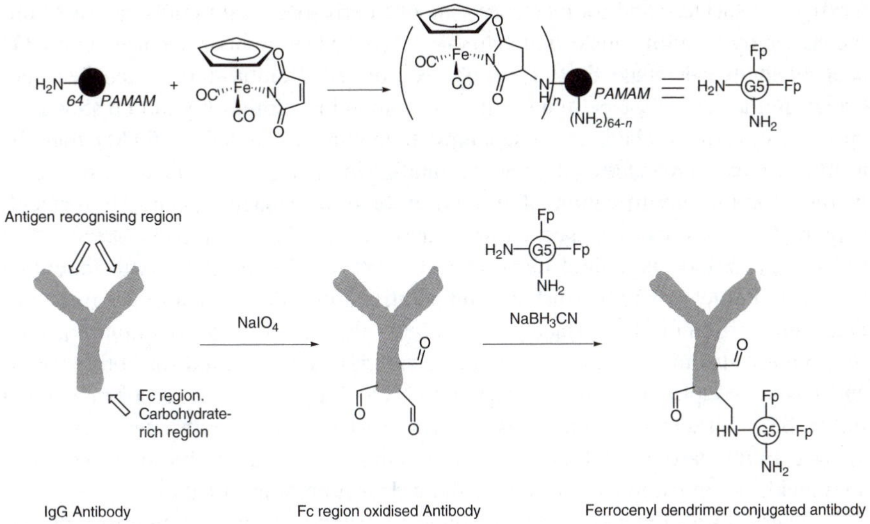

Figure 5.6 *Construction of a dendrimer-labelled antibody for FTIR detection of specific antibody–antigen binding*[30]

The success of this labelling method is claimed to be partly due to the possibility of introducing multiple labels by attaching a few dendrimer molecules only, thereby avoiding an extensive modification of the antibody surface. In fact, in this study, the molar ratio of dendrimer to antibody never exceeded 1.4;[30] the number of Fp units on the dendrimer varied between 10 and 25, the lowest substituted dendrimer reaching the highest coupling to the antibody (ratio: 1.4).

5.3.2 Dendrimers for Amplifying the Covalent Binding Capacity of Solid Phases

The general advantage of assays where a soluble analyte in a sample is detected by its binding to a surface (heterogenous assays) is that the separation of the analyte from other sample components is achieved simply by its binding to the solid phase. If an analyte or another component of such an assay is not optimally adsorbed to the solid phase by passive, non-covalent binding, covalent methods are, however, needed to ensure binding. Covalent attachment methods may also add chemical specificity or selectivity to the binding of the analyte to the solid phase, thereby possibly enhancing the selectivity of the method, or it may enable attachment of biomolecules that are not easily adsorbed by non-covalent methods, an example being carbohydrates or DNA on polystyrene.[31,32] The main challenge of such methods, apart from applying chemical methods with an appropriate reactivity, is to obtain a final, functional capacity that is high enough to achieve a satisfactory level of signal generation. This can be achieved by amplifying the number of reactive groups on the solid-phase surface by derivatisation of dendrimers with the ability of multivalent presentation. Gold surfaces, as used for surface plasmon resonance (a method for investigating receptor–ligand interactions based on mass changes on a reflective surface) and for electrodes may be derivatised with thiols, as for example shown by coupling an Au-reactive thiol PAMAM (Starburst) dendron with a G3 shell of hydroxyl in the first layer followed by a G2 amino-terminated PPI dendrimer in the second layer. When this was coupled to biotin, it could be shown by spectroscopic methods and by scanning tunnelling microscopy (STM) that the biotin groups were accessible for the binding of avidin; thus this constitutes a method for the amplification of the avidin-binding capacity of an Au surface[33] (Figure 5.7). A similar approach was applied by Mark and co-workers[34] who utilised an amino alkyl thiol derivatised G5 PAMAM dendrimer. After coupling various biomolecules to the surface, comparative studies on a "dendrimer free" surface were carried out by surface plasmon resonance. High-stability immobilisation of biomolecules in an easily accessible orientation was achieved for both proteins and DNAs, coupled through primary amines by the use of the homobifunctional linker BS[3] (bis[sulfosuccinimdyl]suberate). Surface plasmon resonance analysis showed an increase of 2.5 times in the immobilisation capacity (biotin probed with streptavidin) compared to a surface without dendrimer amplification.

Similarly, in a model study, an example of a very sensitive immunosensor was demonstrated using dendrimers for the efficient and high-capacity presentation of antigens (in this case, the hapten biotin) on a gold surface in connection with an electrochemical signal-generation system.[34a]

Here, G4-PAMAMs were bound to the gold surface through a succinimide linker and then reacted with biotin to form a covalently coupled avidin, streptavidin and

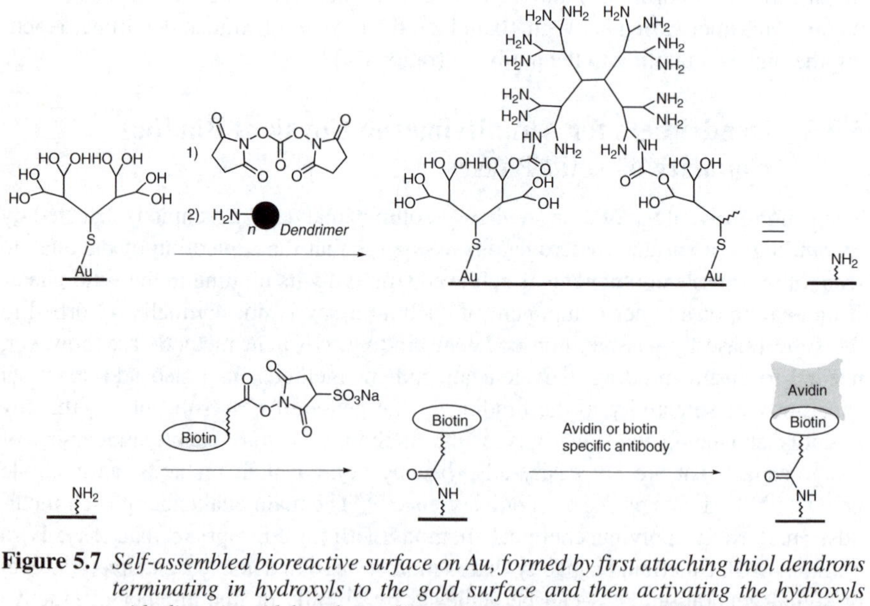

Figure 5.7 *Self-assembled bioreactive surface on Au, formed by first attaching thiol dendrons terminating in hydroxyls to the gold surface and then activating the hydroxyls with disuccinimidyl and coupling amino-terminating dendrimers to this. These dendrimers will both act to crosslink dendrons and to provide multiple amino groups for further coupling reactions, in this case, with biotin. The surface was probed with avidin and investigated by STM[33]*

antibiotin antibody sensing layer. Detection of the bound, peroxidase-labelled molecules was done by electrochemical measurements of peroxidase activity. The same group reported previously that dendrimers (G4 PAMAM) partly derivatised with ferrocenyl groups and biotin groups could be layered onto a gold electrode and could act as a biosensor electrode for avidin, using glucose oxidase and glucose as an electron-generating system.[35] First, a ferrocenyl-coupled dendrimer was coupled to the succinimide-activated gold surface, and second, the coupled dendrimer was derivatised with biotin. Thus, in this setup the dendrimer is thrice heterofunctionalised (gold succinimide, ferrocenyl (around 30% of the surface groups) and biotin). A signal is generated during the interaction with the ferrocenyl groups by the electrons, which are generated by the oxidation of glucose by glucose oxidase. When glucose oxidase is inhibited by the dendrimer-bound avidin to access the electrode surface, the electron transfer from glucose oxidase to the ferrocenyl groups is also inhibited and a decrease in the redox signal from the ferrocene groups will result. This decrease in signal was found to be proportional to the amount of avidin being bound, enabling the quantitation of avidin in the picomolar range (detection limit 4.5 pM). This principle can be extended to any interesting macromolecule binding to the surface of the dendrimer thereby interfering with the electron transfer as also shown with the hapten dinitrophenyl (DNP) and its corresponding antibody.[36] The ability of the ferrocene dendrimer to "sense" glucose in the presence of glucose oxidase was also used to generate a highly sensitive electrode for glucose, and it was shown that the sensitivity and the signal generation increased in proportion to the number of dendrimer and enzyme layers on the electrode[37] (Figure 5.8).

The penicilloylated dendrimers reported by Sanchez-Sancho and co-workers[38] (see also Chapter 6) also exploit the ability of the dendrimer to increase the number of antigens available for reaction with antibodies. In this case, however, the antibodies were of the IgE class characterising allergic reactions and the assay format was the so-called radioallergosorbent test (RAST) in which the putative allergen is adsorbed to a solid phase (here benzylpenicillin coupled to polylysine) and probed with the patient's blood. A positive reaction with IgE as detected by radioactively labelled antiIgE then indicates allergy. The RAST inhibiting potential of penicilloylated G1–G3 PAMAM dendrimers was analysed and compared to a monomeric penicillin. It was shown that the binding of IgE to the penicillin monomer was similar to the different constructs while the molar inhibition capacity increased strongly when going from monomer to dendrimers in the order of G1–G3. This indicates that dendrimers of this class might be used as amplified antigens for *in vitro* allergy tests with an increased sensitivity.

5.3.3 Microarray Application of Dendrimers

Microarray technology allows high-throughput (parallel) analysis of a high number of samples for a big number of genes or gene variants (DNA arrays) or for expression of many different genes after reverse transcription (RT) PCR on an extracted RNA yielding cDNA. However, the necessity of using minute (sub-μL) hybridisation volumes and the very small size of the dots of probe DNA on the microarray slide make it extremely important to apply detection methods with the highest possible analytical

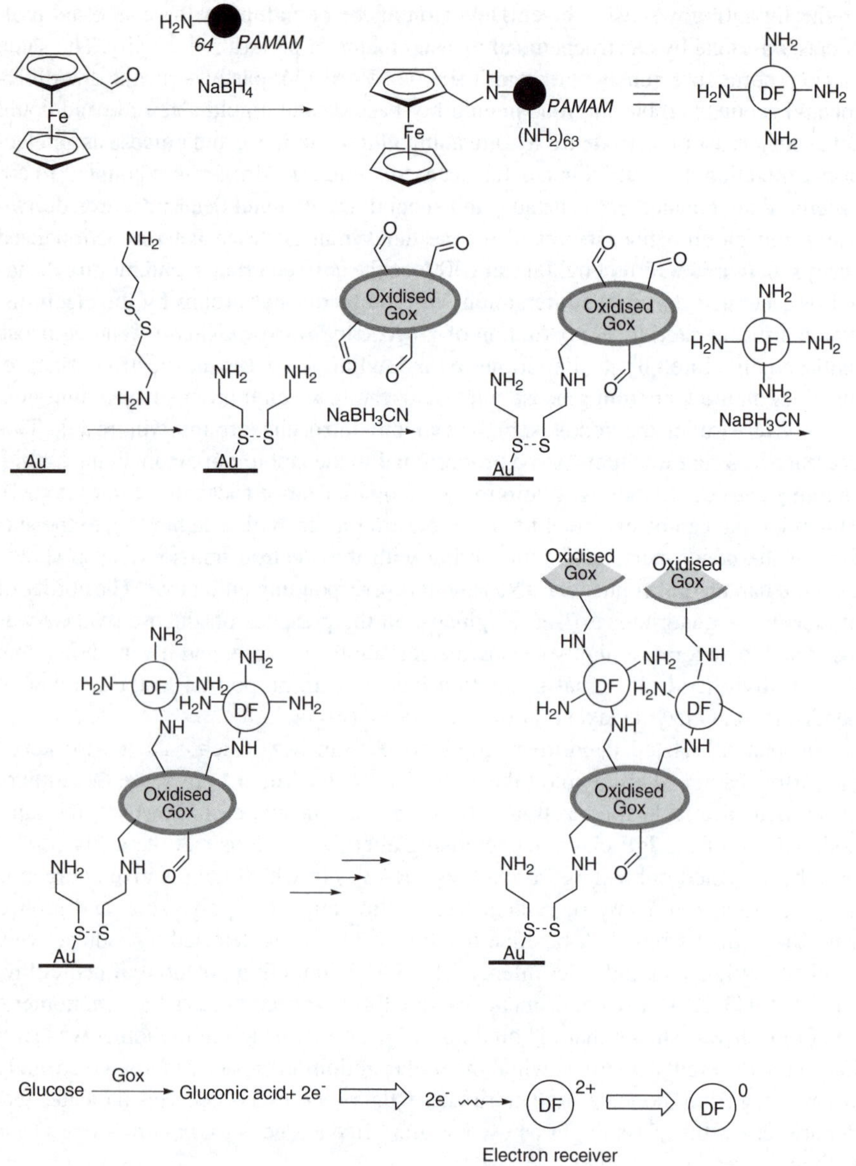

Figure 5.8 *Multilayered electrode consisting of glucose oxidase and dendrimers decorated with ferrocenyl groups as reported by Yoon and co-workers.[37] This electrode can be used for the detection of glucose with high sensitivity. Sensitivity depended on the number of layers and on the degree of substitution of the dendrimer with ferrocenyl groups*

sensitivity. This allows detection of specific nucleic acid sequences at the levels needed for sensitive detection of microorganisms like bacteria and viruses and for sensitive quantitation of eukaryotic DNA and RNA. As for the assays mentioned above, the sensitivity can be increased either by making the hybridisation reactions more efficient, *e.g.* by increasing the number of molecules available for hybridisation in each spot (Figure 5.9) or by employing labelling molecules with enhanced signal-generation capabilities (*vide supra*).

The use of dendrimers to increase the "load" of biomolecules in the spots of a microarray on the most common microarray-based material, glass, has been achieved by aminosilylation followed by activation of the amino groups and coupling of amino-functionalised dendrimers that are then further activated to bind amino groups present in proteins or in aminated oligonucleotides. As an example, 3-aminopropyl-triethoxysilane (or glycidyloxypropyltrimethoxysilane) was used to activate a glass surface followed by coupling of either disuccinimidylglutarate (or glutaric anhydride, followed by *N*-hydroxysuccinimide/*N*,*N'*-dicyclohexylcarbodiimide) and 1,4-phenylenediisothiocyanate and a G5 PAMAM dendrimer which was then activated and partially crosslinked with the same homobifunctional linker for the coupling of amine-containing biomolecules.[39,40] The set up was tested by traditional fluorophore labelled DNA hybridisation and the dendrimer-coated surfaces were found to show a doubling in signal intensity and to be reusable (in contrast to the conventional slides). In addition to DNA, proteins (*e.g.* streptavidin) could also be coupled by this methodology. The system was shown to perform well for single nucleotide polymorphism (SNP) analysis, to be applicable to both oligonucleotide and PCR targets, and furthermore the method resulted in uniform spots on the microarray. Generally, a 10-times increase in signal compared to conventionally coated surfaces was obtained.[39,40]

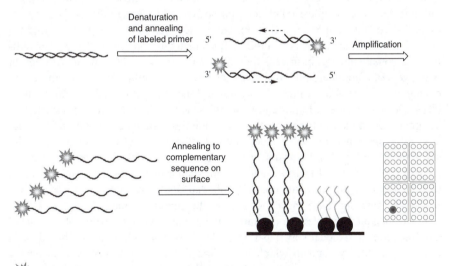

Dendrimer labeled with fluoro genic surface groups

Dendrimer bound to the array surface to increase the number of surface groups

Figure 5.9 *Schematic depiction of the use of dendrimers as fluorescence and surface group density enhancers in microarray technology. In microarrays, the detection agents (specific oligonucleotides, bottom), distributed on discrete "spots" on the array each bind a specific gene. The sample components to be detected (DNA) are labelled, e.g. by PCR with fluorescent primers, a positive reaction being indicated by fluorescence of the spot containing the specific-binding oligonucleotide*

Data on various performance parameters important for optimal microarray assays, including sensitivity and signal:noise ratios have also been reported in the context of various methods for dendrimer-based amplification. For example, cDNA or oligonucleotide probes were coupled non-covalently by electrostatic interactions to G3 DAB dendrimer surfaces (aminosilane-activated glass). UV light stabilised the binding by mediating intermolecular DNA crosslinking.[41] In this case, dendrimers did not give any advantage, but covalently coupled DNA was found to lead to a considerable increase in sensitivity,[42] allowing the use of minute amounts of probe material in the spots and resulting in an increased detection sensitivity. The latter dendrimers were based on a hexachlorocyclotriphosphazene core and G4 dendrimers terminated with aldehyde functions were used, coupling 5'-amine-modified DNA by reductive amination.

For antibody arrays, in which antibody molecules are bound by dendrimer epoxy groups, a considerable gain in signal with fluorescently labelled antigen was obtained with no increase in the background signal, increasing the detection limit 2–20 times depending on the type of slide it was compared to. The dendrimer type was not disclosed. When a protein antigen (human serum albumin) was coupled to this surface, and subsequently detected with fluorescent antibody, the increase in sensitivity was not as pronounced, however the dendrimer slides, together with one brand of amine slides showed the lowest detection limit of the surfaces tested.[43]

The use of dendrimers as enhancers of signal:noise ratios has been demonstrated in a DNA microarray designed to detect a number of viruses.[20] Small phosphoramidite dendrons, synthesised by solid phase synthesis were used to conjugate two, three, four or nine copies of a fluorophore to the 5' end of DNA primers used for amplification of the viral DNA in the samples being hybridised on the array, resulting in a considerable gain in sensitivity. The primers were derivatised with one or two layers of a commercially available "doubler" or a "trebler" phosphoramidite to contain two, three, four or nine 5'-alkylamino functions onto which the fluorophore NHS ester was coupled. Importantly, the efficiency of the PCR was not affected by incorporation of the fluorophore-dendrimer moieties compared to the normal single-labelled primers and the subsequent microarray reactions also retained specificity and reproducibility. The highest sensitivity increase was observed for low levels of signal (quenched signal DNA) where the nine-dendrimer derivatised DNA gave up to 30 times the signal seen with normally labelled DNA. The three-dendrimer derivatised DNA gave signals in the same range as the nine-dendrimer.

In an elegant example demonstrating the versatility of dendrimer-based label amplification, a radioactively labelled (^{32}P) oligonucleotide dendron based on a phosphoramidite synthon in which the radioactive label was incorporated on the 5'ends by enzymatic means was investigated.[44] This dendron was combined with a specific, hybridising oligonucleotide in a dendron head-monomeric oligonucleotide tail arrangement yielding a polylabelled DNA hybridisation probe. Furthermore, an oligonucleotide–dendron based on the same principles could be used as a primer for PCR amplification of a target DNA sequence yielding highly labelled products.

Another type of all-DNA dendrimer, described above (also see Chapter 1), is assembled through inter-hybridising oligonucleotide building blocks (monomers).[23] The use of this dendrimer for amplifying radioactive signals, enabling sensitive detection of the widespread human herpes virus, Epstein–Barr virus, by detection of its specific RNA in peripheral blood mononuclear cells has been described.[45] The dendrimers were labelled by the annealing of radiolabelled oligonucleotides to dendrimer surface oligonucleotides with complementary sequences while the other surface oligonucleotides of the dendrimer were designed to anneal to viral RNA extracted from the infected cells and blotted onto a membrane as a model of a microarray design. The sensitivity reached 10% of that of PCR. This DNA–dendrimer type was also used with fluorescent labelling in the following way: the two available types of free oligonucleotide strands of the DNA dendrimer were used to bind a flourophore-labelled DNA strand (>200 binding strands available on the surface of the dendrimer) and a 5′-oligonucleotide extension of RT-PCR-generated cDNA samples, respectively.[46] Thus, the dendrimer effectively bridges the cDNA bound to the microarray and the fluorophore. The signals obtained correlated nicely to the amount of cDNA in the sample, kept the background very low and increased the sensitivity for mRNA detection 16 times compared to a conventional cDNA microarray.

5.3.4 Fluorophore-Labelled Dendrimers for Visualisation Purposes

Fluorescent G5-PAMAM dendrimers were applied for visualising folate receptor-expressing tumours[47] (*vide supra*) using a bi-dendrimer construct bound together by complementary DNA strands on the two dendrons, one dendrimer carrying, in addition, folic acid and the other dendrimer labelled with fluorescein. Labelling of and uptake by folate receptor expressing cells were demonstrated. The whole bi-dendrimer complex had a diameter of 20 nm with 11 nm being constituted by the DNA spacer/linker and five fluorescein molecules were present on the first dendrimer and two folic acid moieties on the second one. No selective chemistry was employed; ratios of fluorescein and folic acid labelling were controlled by the stoichiometry of the reactions. The DNA-binding step was not quantitative and there were problems with entrapment of fluoresceine isothiocyanate inside the dendrimer. Despite these problems, the principle opens up for a range of both diagnostic and therapeutic reagents combining different targeting dendrons with different visualising og drug-carrying dendrons (see Chapters 1 and 4).

Although self-quenching can be utilised in fluorescent assays for protease activity[22,48] (*vide infra*), this is a problem with dendrimer amplified fluorescence which seeks to enhance the fluorescence of several fluorophore moieties coupled to one dendrimer. For fluorophores with small Stoke's shifts (*e.g.* fluorescein) a close positioning on the dendrimer surface will lead to an intramolecular self-quenching of fluorescence, as was for example, demonstrated with a series of fluorescein-coupled dendrons going from two to six "arms" where the emission essentially decreased to zero for the six-mer construct.[48] In a different design, using non-self-quenching fluorophores, Wang and co-workers[49] investigated G1- to G5-PAMAM dendrimers derivatised with either phenylenefluorene or phenylenebis(fluorene). These dendrimers are cationic and fully water soluble. FRET could be demonstrated using such fluorescent dendrimers as donors and fluorescein-labelled double-stranded (ds)

DNA (16-mer) as acceptors and the dendrimer thus acted as a water soluble light-harvesting dendrimer able to enhance an incoming light signal and transfer it to a oligonucleotide bound acceptor or reporter molecule, resulting in a considerable increase in the emission intensity of the reporter (Figure 5.10). With increasing generations, very large molar absorption coefficients are reached (up to 18) and very efficient transfer of energy takes place between the dendrimer fluorophore and the fluorescein-coupled dsDNA due to the multivalent interactions between the cationic dendrimer and the anionic dsDNA. No FRET took place with free fluorescein. However, reaching the higher generations, a lowering of the quantum efficiencies and the extinction coefficient per optical unit is seen, probably due to crowding of the surface fluorophore units, leading to a plateau in fluorescein emission intensity.[49]

Protease activity can be monitored by recording the increases in fluorescence due to the separation of a quencher moiety from a reporter moiety by the action of the protease in question on an adequately labelled peptide; this can be amplified by use of dendrimer labels, *e.g.* by placing the peptide in question, labelled N terminally with a self-quenching fluorophore in several copies in a parallel arrangement attached to dendrimer surface groups. This distance is adequately close to allow for an efficient quenching (low background) which is abolished upon cleavage of the peptide.[22,48] Ellard and co-workers[48] investigated trivalent and divalent dendrons coupled to an octameric peptide N-terminally labelled with the fluorophore Cy5 (synthetic sulfoindocyanine dye). In addition, a tetrameric peptide labelled with fluorescein was coupled to divalent and trivalent dendrons. A large increase in fluorescence (up to nine times) were observed upon protease-mediated cleavage of these

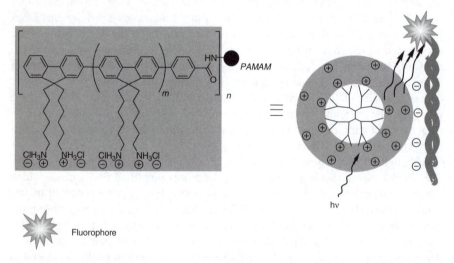

Figure 5.10 *Light-harvesting, water-soluble dendrimer of Wang and co-workers.[49] PAMAM G1–G5 dendrimers were surface functionalised with phenylenefluorene or phenylenebis(fluorene) groups and treated with an acid, yielding positively charged, water soluble fluorescent dendrimers that could act as a FRET donor to a fluorescein acceptor attached to a model 16-mer DNA oligonucleotide. The DNA is binding to the dendrimer through electrostatic forces*

peptide substrates. This system has the advantage of only one type of fluorophore being needed, allowing simple synthesis routes to be employed. However, when applying this method for peptide probing of proteases, molecular crowding of the peptides might become a problem. Problems due to crowding have consistently been observed with MAP peptides having a similar arrangement of the peptide units (see Chapter 4). Another way is to use long-reaching fluorophore-quencher pairs at both ends of the peptide substrate, using a dendrimer as scaffold for multivalent presentation of fluorophores (should not be self-quenching). Upon cleavage of each peptide molecule, multiple fluorophores become dequenched and accordingly an enhanced fluorescence signal is obtained (Figure 5.11).[22] These are versatile approaches and can be used to label multiple peptides for peptide library screening of peptide substrates for proteolytic enzymes.

5.3.5 Using Dendrimers to Increase DNA Extraction Yields

A simple way to enhance the sensitivity of DNA-detection methods is to increase the yield of DNA extracted from the sample for subsequent analysis. This could be accomplished through the use of dendrimers in the extraction step[50] as an example of surface derivatisation of dendrimers for preparative purposes. Polyamidoamine dendrons (up to G6) were synthesised on the surface of magnetite particles through an aminosilane bond and the enhanced density of the positively charged amino

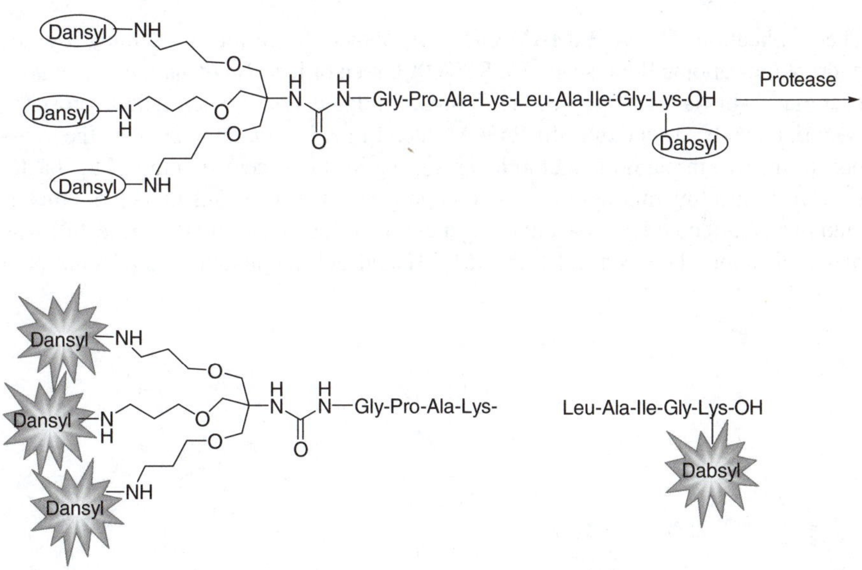

Figure 5.11 *A design for "liberation" of fluorescence by the action of a peptide-cleaving protease, utilising a dendron-like arrangement of the fluorescent moieties;[22] the uncleaved peptide allows a long-range quenching to take place between the fluorophore at the C-terminal of the peptide (dabsyl) and the three fluorescent dansyl molecules attached to the N-terminal. Quenching is terminated upon proteolytic cleavage of the peptide*

groups was used to achieve an improved yield of DNA extracted from solutions of DNA by polyplex formation (see Chapters 2 and 3). The magnetite particles had diameters in the range of 50–100 nm and benefitted from the cationic dendrimer coating by becoming increasingly dispersed in the suspension. This dispersion increased the accessible number of cationic groups able to participate in the binding of DNA. The number of accessible amino groups increased linearly up to G5. Increasing the size to a G6 dendrimer did not increase the amino groups linearly due to steric interference of the sulfo-LC-SPDP used to monitor the accessibility of surface amines; however, the increased dispersion of particles obtained with high-generation dendrimers increased the efficiency of the adsorbent so that the coupling capacity for DNA was doubled from G5 to G6. By contrast, DNA binding capacities of low generation particles increased only slightly when increasing the generation, due to the aggregation of the particles occurring at these generations.

Dendrimers (G3 PAMAM) were also used by Lei and co-workers[51] to prepare a high-capacity resin with small (3–4 μm), non-porous zirconium and urea formalde-hyde co-polymer particles for high-performance liquid chromatography of various biological macromolecules and nucleotides, using RNA as the stationary phase. The PAMAM dendrons were synthesised directly on the imido groups of the resin and the RNA was coupled to the amino groups of the dendrimers using RNA brominated in the position eight in the purine nucleoside base (Figure 5.12).

5.3.6 Using Dendrimers for direct Detection of Live Bacteria

The application of G4-PAMAM-OH was shown to result in staining by the hydrophilic cationic fluorescent dye SYTOX Green of live *Pseudomonas aeruginosa* bacteria.[52] This is interesting as the fluorescent dye is not by itself able to cross the bacterial membrane and thus the PAMAM dendrimer is thought to mediate the trans-port of the dye through the membrane resulting in fluorescent staining of the bacte-rial DNA just by mixing the two components. The mechanism behind this is unknown, but it could involve either a general increase in membrane permeability by hydrogen bonds between G4-PAMAM-OH and cell-membrane components or it

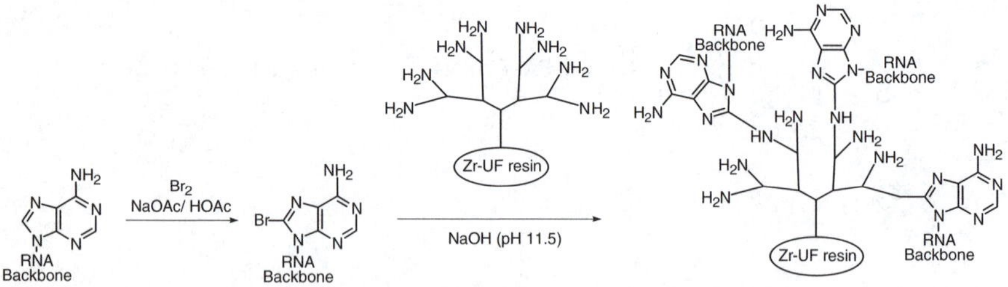

Figure 5.12 *Zirconium-based resin coupled with brominated RNA.[51] The base resin was acti-vated with urea and formaldehyde and then derivatised with methyl acrylate and ethylene diamine by on-resin synthesis to generate G-3 dendrimers. The result-ing 3–5 μm, non-porous beads were used for high-performance liquid affinity chromatography of biological macromolecules*

could involve formation by electrostatic and hydrophobic bonds of a SYTOX Green–PAMAM-OH complex that had membrane penetrating properties.

The fluorescence intensity correlated with the concentration of the bacteria and the dendrimer stabilised the dye, making it possible to manufacture a dried film from these components for the rapid and easy detection of bacteria applied as a droplet suspension to the film and read by a fiber optics-based fluorometer.

5.4 Summary

In conclusion, dendrimers have broad applicability in diagnostic methods, including their use as contrast agents for imaging and in various bioassays. Dendrimers are useful imaging agents in proton-, X-ray- and radioactivity-based methods as they can be designed to carry ionic contrast elements by chelation and others by covalent binding and thereby provide a shielding from the toxicity of these components. In addition, the dendrimer binds multiple contrast yielding elements, creating a powerful contrast agent, and by manipulating the dendrimer size and shape, different biodistributions and retention times can be obtained, for example, for imaging of soft tissue by a vascular distribution of the agent. There is also a possibility of specific targeting by including a specific ligand in the dendrimer-contrast agent. Most work has been done with the proton-based reagents, which work at far lower concentrations than X-ray reagents and which may be safer than radioactivity-based compounds.

Dendrimers have also been investigated for use in bioassays, especially for the enhancement of sensitivities in microarray and ELISA techniques. Commercial reagents based on all-DNA dendrimers[23] for general, highly amplified labelling of PCR products for microarray detection are already available. Even if this type of dendrimer is generally not molecularly perfect, its versatility and proven amplification potential make it a very useful tool for signal enhancement in DNA-based assays. Dendrimers can also be used to enhance sensitivity by amplifying the covalent-binding capacity of solid surfaces, however, problems of increased steric crowding of biological macromolecules attached to the surface groups of higher generation dendrimers have precluded a wide use of this principle. The most obvious uses will comprise low-generation dendrimers and small molecules, (*e.g.* haptens) an example being low-generation dendrons used for fluorescence-based protease assays on peptide libraries.

In most of the bioassay applications described here, there are no specific benefits to be gained from the homogenicity and the compactness of a molecular flawless dendrimer. Instead, it is the multitude of derivatisable surface groups and their chemical versatility that make dendrimers interesting in the bioassay context and they therefore have to compete with a number of other types of polymers, including well-known, biocompatible and inexpensive polymers like dextran, PEG, *etc.*

References

1. P.M. Conn (ed), *Meth. Enzym.*, 2004, **385**, 3.
2. P.M. Conn (ed), *Meth. Enzym.*, 2004, **386**, 3–418.
3. W. Krause, N. Hackmann-Schlichter, F.K. Maier and R. Müller, *Top. Curr. Chem.*, 2000, **210**, 261.

4. W. Krause (ed), *Top. Curr. Chem.*, 2002, **221**, 1.
5. W. Krause (ed), *Top. Curr. Chem.*, 2002, **222**, 1.
6. A.T. Yordanov, A.L. Lodder, E.K. Woller, M.J. Cloninger, N. Patronas, D. Milenic and M.W. Brechbiel, *Nano Lett.*, 2002, **2**, 595.
7. H. Schumann, B.C. Wassermann, S. Schutte, J. Velder, Y. Aksu, W. Krause and B. Raduechel, *Organometallics*, 2003, **22**, 2034.
8. H. Suzuki, H. Kurata and Y. Matano, *Chem. Commun.*, 1997, 2295.
9. K.E. Appel, *Drug Metabol. Rev.*, 2004, **36**, 786.
10. T. Klapotke, *Biol. Met.*, 1988, **1**, 69.
11. E.C. Wiener, F.P. Auteri, J.W. Chen, M.W. Brechbiel, O.A. Gansow, D.S. Schneider, R.L. Belford, R.B. Clarkson and P.C. Lauterbur, *J. Am. Chem. Soc.*, 1996, **118**, 7774.
12. E.C. Wiener, M.W. Brechbiel, H. Brothers, R.L. Magin, O.A. Gansow, D.A. Tomalia and P.C. Lauterbur, *Mag. Res. Med.*, 1994, **31**, 1.
13. M.W. Bourne, L. Margerun, N. Hylton, B. Campion, J.J. Lai, N. Derugin and C.B. Higgins, *J. Mag. Res. Imag.*, 1996, **6**, 305.
14. H. Kobayashi, S. Kawamoto, S.-K. Jo, L.H. Bryant, M.W. Brechbiel and R.A. Star, *Bioconj. Chem.*, 2003, **14**, 388.
15. N. Sato, H. Kobayashi, A. Hiraga, T. Saga, K. Togashi, J. Konishi and M.W. Brechbiel, *Mag. Res. Med.*, 2001, **46**, 1169.
16. S.J. Wang, M. Brechbiel and E.C. Wiener, *Invest. Radiol.*, 2003, **38**, 662.
17. S.D. Konda, M. Aref, S. Wang, M. Brechbiel and E.C. Wiener, *Magma (New York, NY)*, 2001, **12**, 104.
18. E.C. Wiener, S. Konda, A. Shadron, M. Brechbiel and O. Gansow, *Invest. Radiol.*, 1997, **32**, 748.
19. M. Subbarayan, S.J. Shetty, T.S. Srivastava, O.P.D. Noronha, A.M. Samuel and H. Mukhtar, *Biochem. Biophys. Res. Commun.*, 2001, **281**, 32.
20. H.-M. Striebel, E. Birch-Hirschfeld, R. Egerer, Z. Földes-Papp, G.P. Tilz and A. Stelzner, *Exp. Mol. Pathol.*, 2004, **77**, 89.
21. J.O. McIntyre, B. Fingleton, K.S. Wells, D.W. Piston, C.C. Lynch, S. Gautam, and L.M. Matrisian, *Biochem. J.*, 2004, **377**, 617.
22. M. Ternon, J.J. Díaz-Mochón, A. Belsom and M. Bradley, *Tetrahedron*, 2004, **60**, 8721.
23. T.W Nilsen, J. Grayzel and W. Prensky, *J. Theor. Biol.*, 1997, **187**, 273.
24. S. Capaldi, R.C. Getts and S.D. Jayasena, *Nucleic Acids Res.*, 2000, **28**(7), e21.
25. www.genisphere.com.
26. M. Lowe, A. Spiro, Y.-Z. Zhang and R. Getts, *Cytometry*, 2004, **60A**, 135.
27. M.K. Borucki, J. Reynolds, D.R. Call, T.J. Ward, B. Page and J. Kadushin, *J. Clin. Microbiol.*, 2005, **43**, 3255.
28. Y. Li, Y.T.H. Cu and D. Luo, *Nat. Biotech.*, 2005, **23**, 885.
29. T. Horn, C.-A. Chang and M.S. Urdea, *Nucleic Acids Res.*, 1997, **25**, 4842.
30. N. Fischer-Durand, M. Salmain, B. Rudolf, A. Vessières, J. Zakrzewski and G. Jaouen, *ChemBioChem*, 2005, **5**, 519.
31. E.S. Jauho, U. Boas, C. Wiuff, K. Wredstrom, B. Pedersen, L.O. Andresen, P.M.H. Heegaard and M.H. Jakobsen, *J. Immunol. Meth.*, 2000, **242**, 133.

32. T. Koch, N. Jacobsen, J. Fensholdt, U. Boas, M. Fenger and M.H. Jakobsen, *Bioconj. Chem.*, 2000, **11**, 474.
33. M. Yang, E.M.W. Tsang, Y.A. Wang, X. Peng and H.-Z. Yu, *Langmuir*, 2005, **21**, 1858.
34. S.S. Mark, N. Sandhyarani, C. Zhu, C. Campagnolo and C.A. Batt, *Langmuir*, 2004, **20**, 6808.
34a. H.C. Yoon, H. Yang and Y.T. Kim, *Analyst*, 2002, **127**, 1082.
35. H.C. Yoon, M.-Y. Hong and H.-S. Kim, *Anal. Biochem.*, 2000, **282**, 121.
36. B.Y. Won, H.G. Choi, K.H. Kim, S.Y. Byun, H.S. Kim and H.C. Yoon, *Biotechnol. Bioeng.*, 2005, **89**, 815.
37. H.C. Yoon, M.-Y. Hong and H.-S. Kim, *Anal. Chem.*, 2000, **72**, 4420.
38. F. Sanchez-Sancho, E. Perez-Inestrosa, R. Suau, C. Mayorga, M.J. Torres and M. Blanca, *Bioconj. Chem.*, 2002, **13**, 647.
39. R. Benters, C.M. Niemeyer and D. Wöhrle, *Chembiochem*, 2001, **2**, 686.
40. R. Benters, C.M. Niemeyer, D. Drutschmann, D. Blohm and D. Wöhrle, *Nucleic Acids Res.*, 2002, **30**(2), e10.
41. S. Taylor, S. Smith, B. Windle and A. Guiseppe-Elie, *Nucleic Acids Res.*, 2003, **31**(16), e87.
42. V. Le Berre, E. Trévisiol, A. Dagkessamanskaia, S. Sokol, A.-M. Caminade, J.P. Majoral, B. Meunier and J. Francois, *Nucleic Acids Res.*, 2003, **31**(16), e88.
43. P. Angenendt, J. Glökler, J. Sobek, H. Lehrach and D.J. Cahill, *J. Chromatogr. A*, 2003, **1009**, 97.
44. M.S. Shchepinov, I.A. Udalova, A.J. Bridgman and E.M. Southern, *Nucleic Acids Res.*, 1997, **25**, 4447.
45. R.J. Orentas, S.J. Rospkopf, J.T. Casper, R.C. Getts and T.W. Nilsen, *J. Virol. Meth.*, 1999, **77**, 153.
46. R.L. Stears, R.C. Getts and S.R. Gullans, *Physiol. Genomics*, 2000, **3**, 93.
47. Y. Choi, T. Thomas, A. Kotlyar, M.T. Islam and J.R. Baker, *Chem. Biol.*, 2005, **12**, 35.
48. J.M. Ellard, T. Zollitsch, W.J. Cummins, A.L. Hamilton and M. Bradley, *Angew. Chem. Int. Ed.*, 2002, **41**, 3233.
49. S. Wang, J.W. Hong and G.C. Bazan, *Org. Lett.*, 2005, **7**, 1907.
50. B. Yoza, A. Arakaki and T. Matsunaga, *J. Biotech.*, 2003, **101**, 219.
51. S. Lei, S. Yu and C. Zhao, *J. Chromatogr. Sci.*, 2001, **39**, 280.
52. A.-C. Chang, J.B. Gillespie and M.B. Tabacco, *Anal. Biochem.*, 2001, **73**, 467.

CHAPTER 6

Dendrimers as Biomimics

6.1 Introduction

Dendrimers with their macromolecular dimensions and well-defined and compartmentalised structure are ideal mimics for a wide variety of biomolecules. The flexible design of dendrimers opens up for the possibility to create microenvironments, *e.g.* hydrophobic or hydrophilic pockets surrounding the active site of dendrimeric artificial hosts or enzymes. The commercially available dendrimer types provide microenvironments with highly diverse properties. PAMAM dendrimers with their network consisting of numerous mixed tertiary amines and secondary amides, create a polar environment capable of, *e.g.* base catalysis (via the amines) and/or hydrogen bond donation (amides) and acceptance (amides and the amines), which make these dendrimers useful as hosts for polar acidic guests and in catalysis of chemical reactions involving polar reactants. Fréchet-type dendrimers built up by numerous of alkoxybenzylether moieties, an environment interact with reactants/guests with aromatic stucture, *e.g.* by π-interactions, and in addition, the π-system may stabilise cations by electron donation. In addition, the oxygen atoms present in the Fréchet dendrimers have the ability to interact with polar substrates/guests by electron pair donation, *e.g.* as hydrogen bond acceptors. The multivalent surface may be, as we have already seen, modified with various molecular motifs, for the formation of, *e.g.* antigens or other biopolymers having highly repeating structures. Dendrimers thus provide ideal molecular motifs for the mimicking of functional biomolecules as they have

- the ability to expose a multivalent surface, for increased binding of biomolecules, otherwise binding weakly to their natural substrate/ligand, useful in mimicking, *e.g.* an artificial cell surface,
- the ability to create a microenvironment inside the dendrimer, giving possibility to create artificial catalytic sites, or cavities having properties different from the surroundings, *e.g.* hydrophobic, hydrophilic cavities, *etc.* useful for the construction of enzyme mimics; and
- the ability to provide a well-defined spatial arrangement of functional subunits, with a high molecular density compared to natural systems, leading to enhanced stabilisation of the three-dimensional structures.

The following chapter will give a survey of the attempts to create functional mimics of natural systems by applying dendrimers as scaffolds or encapsulation entities. Biomimicking behaviour is also used, *e.g.* in the development of drugs and drug delivery devices, where the dendrimeric adducts are utilised, *e.g.* as mimics of cell-surfaces. As the preceding chapters have treated this subject in greater detail, this part of the biomimicry area will only be briefly mentioned in this chapter.

6.2 Dendrimers as Protein Mimics

The molecular dimensions and globular conformation and compartmentalised structure of high-generation dendrimers, already early in the history of dendrimers led researchers to the idea of utilising these synthetic macromolecules as mimics of globular proteins.[1,2] As seen in Chapter 1, the dendrimeric structures respond to changes in the surroundings, *e.g.* solvent polarity, ionic strength, *etc.*, and may depending on these conditions expose exterior or interior structures with different properties. However, the more dense and cross-linked structure of dendrimers give these systems more favourable qualities compared to naturally occuring proteins.

As a result of their more densely packed structure compared to natural proteins (see Chapter 1), certain peptide-based dendrimer systems (*e.g.* MAP systems) show a significantly increased resistance towards proteases.[3] These dense protease resistant-peptide dendrimeric structures very well mimic densely aggregated protein structures sometimes observed in nature. In natural systems, protease resistance indicates excessive formation of protein aggregates, a critical state observed in, *e.g.* amyloid fibrillar diseases like Creutzfeldt Jakob's or Alzheimer's disease. Dendrimers with a partly modified outer shell and a reactive core resulting in a more open surface structure compared to a completely surface modified dendrimers, are found to be more prone to form "megamers" (dendrimer aggregates), whereas perfectly shaped dendrimers with a more closed shell do not show a likewise high tendency to aggregate.[4]

Dendrimer-based collagen mimics have recently been synthesised by Goodman's group.[5] In these constructs, the denrimer is used as a building block to mimic a non-globular collagen structure, showing that dendrimers, although being mostly globular shaped, may be used as mimics of non-globular structures. These collagen mimics were based on a trimesic acid-cored dendrimer carrying three triple arm forks decorated with the Gly-Nleu-Pro collagen mimetic sequence (Nleu = norleucine). The three "triple forks" of peptide repeats form three triple-helical structures emerging from the trimesic core. These collagen mimics showed higher triple-helical stability in comparison to similar constructs based on non-dendritic scaffolds, and the higher stability was found to be independent on dendrimer concentration, thus being result of intramolecular stabilisation of the triple helical peptide clusters. It is envisioned that these dendrimer-based collagen mimics may serve as ideal scaffolds for further functionalisation to artificial enzymes, drugs or drug delivery devices or other biomedical applications.

Besides mimicking the building blocks for tissue formation, dendrimers may also mimic the numerous of protein-based receptors utilised in nature for specific biological recognition.

Dendroclefts, a class of dendrimers (see Chapter 3), which are highly stereospecific in recognising and binding of carbohydrates, have the ability to differentiate between the anomeric forms of glucose, preferentially binding the β-anomer (Figure 6.1).

The dendroclefts show a generation-dependent behaviour in guest binding, where the more "closed" high generation structure increases the interaction between the PEG oxygen atoms and the carbohydrate, increasing the stereospecificity. With their specific carbohydrate binding properties, the dendroclefts mimics the function of lectins, which constitutes an important class of naturally occuring proteins involved in carbohydrate-mediated signalling and binding.[6]

An alternative way of utilising dendrimers in the creation of artificial receptors, *i.e.* mimicking the biological recognition processes was published recently by Zimmerman's group, who used dendrimers as "monomers" for molecular imprinting in a supramolecular polydendrimeric network. In molecular imprinting, the polymerisation takes place in the presence of a substrate, which is subsequently selectively removed to create a polymer with built-in substrate cavities (Figure 6.2).[7] Such imprinted polymers have been extensively investigated as artificial receptors, thereby mimicking the biological recognition normally performed by proteins, *e.g.* antibody-antigen or enzyme–substrate recognition.[8] In the dendrimeric approach dendrimers which contain a substrate (porphyrin) core and polymerisable surface groups are polymerised creating a "poly-dendrimer" polymer containing numerous

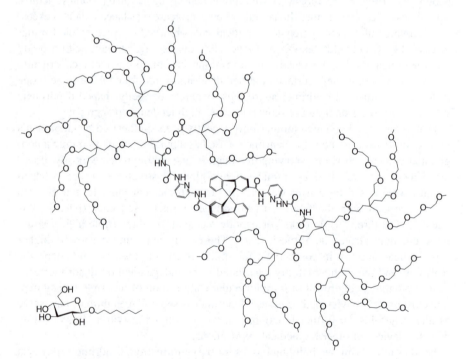

Figure 6.1 *The structure of a G2-dendrocleft. The dendrocleft mimics the stereospecific recognition between lectins and carbohydrates by preferential binding of the β-anomeric glycosyl derivative[6]*

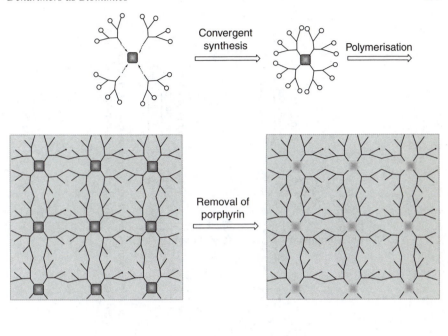

Figure 6.2 caption with text:

Convergent synthesis

Polymerisation

Removal of porphyrin

■ = Porphyrin ◯ = Polymerisable group

Figure 6.2 *Two-dimensional schematic depiction of the use of dendrimers as monomers in the formation of molecular imprinted supramolecular polymers[7]*

porphyrins. The porphyrins are cleaved selectively from the cross-linked dendrimers by hydrolysis of base labile ester bonds, forming multiple porphyrin cavities. The resulting poly-dendrimer polymer was able to selectively recognise porphyrin with a binding constant of $1.4 \times 10^5 \text{ M}^{-1}$. In comparison to molecular imprinting based on traditional monomers, the use of dendrimers as monomer is advantageous because of the ability to preform the monomeric dendritic hosts in a spatially well-defined manner. The structural integrity of substrate cavities in the poly-dendrimer skeleton should therefore be more uniform throughout the polymer, compared to traditional imprinted polymers.

Dendrimers mimicking the protein endostatin have been synthesised and investigated as angiogenesis-inhibitors in cancer therapy.[9] Endostatin, a small protein derived from the C-terminal domain of collagen XVIII has shown good ability in inhibiting the development of blood vessels (angiogenesis) in tumour tissues, disabling the tumour metastasis growth (Figure 6.3). Endostatin has a cationic arginine-rich domain, which binds negatively charged heparins and heparan-sulfates resident on the vascular walls. These negatively charged carbohydrates act as stimulatory co-factors in the binding of the "vascular endothelial growth factor" (VEGF) to receptors at the vascular wall. The simultaneous dual binding of VEGF to heparin and VEGF receptors at the vascular endothelia is a crucial step in the initiation of vascular development (Figure 6.3). To ensure high metastasis and growth rate, VEGF is over-expressed in cancer tumours.

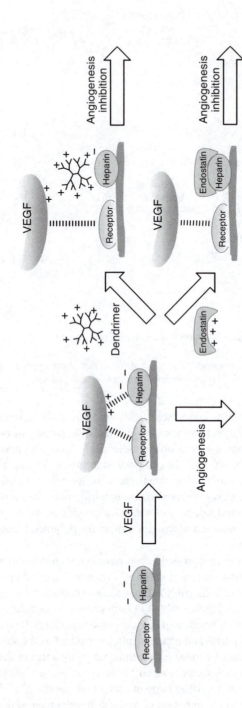

Figure 6.3 *Depiction of a proposed mechanism for angiogenesis inhibition by (top) cationic arginine-containing dendrimers (endostatin mimics) or endostatin (bottom), competing with VEGF for binding to negatively charged heparin on the vascular wall, presumably via electrostatic interactions*

As endostatin mimics, G2- and G3-polylysine dendrimers were synsthesised with multiple terminal arginine residues, thereby providing highly multivalent cationic surfaces. In addition to G2- and G3-arginine-containing dendrimers, citrulline building blocks were utilised to create a more neutral urea-containing dendrimer surface. The ability to bind heparin was found to be generation dependent as well as charge dependent, with the G3-arginine dendrimer having the highest affinity for heparin, and the urea-containing citrulline dendrimer having the lowest heparin affinity. Also, in the test of antiangiogenesis properties, *e.g.* by CAM assay (chicken embryo chorioallantoic membrane assay – the inhibition of blood vessel formation by measuring the formation of an avascular zone inside the egg) the arginine-containing dendrimers showed high anti-angiogenetic properties similar to endostatin and angiostatin (heparin-binding proteins).

The citrulline dendrimers showed low anti-angiogenetic activity, however, significant inhibition of angiogenesis was obtained at higher concentrations of citrulline dendrimer, indicating that the presence of N-terminal protonised primary amines on the dendrimer surface also have anti-angiogenetic properties similar to the guanidinium groups on the arginine-containing dendrimers.

In addition to the use of proteins as "passive" building blocks in the build up of skin- and tissue barriers or other biomaterials, proteins actively take part in a large variety of biological processes, and mimicry of the specific properties of these molecules is an interesting task to get a deeper understanding of the intrinsic molecular factors involved in these processes. Important functions of proteins in living systems include the transport of electrons, *e.g.* in the cytochrome electron-transport systems, or the transport of low-molecular species such as molecular oxygen. Both in the electron transport and in the transport of molecular oxygen, haem-containing proteins constitute an important group of proteins.

Dendrimeric electron transporters: In cytochromes, long-distance electron transfer, from the haem-group metal centres through the protein architecture towards electron acceptors in the "electron transport chain" is a process strongly dependent on the site isolation of the redox centre from the surroundings, thus similar electron transfer reactions are not observed in unprotected haem-systems. As the conformations of higher generation dendrimers provide an interior environment with strongly divergent properties from the solvent, these high-generation dendrimers should serve as good artificial hosts for long-range electron transfer processes, not plausible with unprotected haem-groups.[10]

Diederich's group found that in dendritic cytochrome mimics the extent of dendritic branching, *i.e.* the dendrimer generation greatly influences the redox potential of the Fe III/FeII couple. In apolar solvents like dichloromethane an increase in the FeIII/FeII redox potential from -0.21 V to $+0.10$ V was observed going from G0 to G1, where the redox potential of the G0 dendrimer resembles that of a bis-imidazole ligand iron-porphyrin complex. The higher redox potential of the G1 adduct indicates a strong shielding of the porphyrin system from the solvent by the dendritic wedges (Figure 6.4).

When applying water as a solvent, a drastic change in redox potential was observed when going to G2 adduct, showing that water because of its higher polarity interacts more strongly with the porphyrin-iron complex, and hence a higher degree of shielding

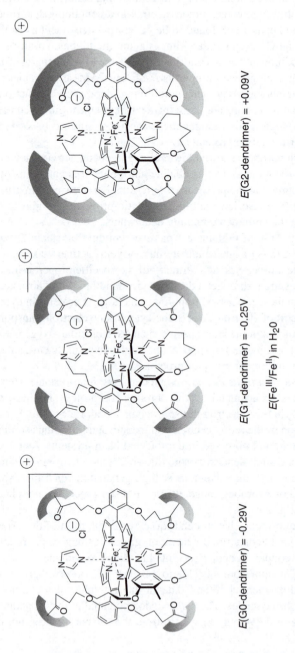

E(G0-dendrimer) = -0.29V E(G1-dendrimer) = -0.25V E(G2-dendrimer) = +0.09V

E(Fe^{III}/Fe^{II}) in H$_2$O

Figure 6.4 *Cytochrome b5 mimic by Weyermann and co-workers.[10] The increased shielding by the dendritic wedges (depicted as grey arches) leads to an increased redox potential, which is independent of solvent conditions, similar to what is observed in natural systems*

is needed to create a microenvironment. Interestingly, the redox-potential of the G2 adducts in dichloromethane and water are in the same magnitude, whereas larger differences between the potentials in water and dichloromethane were observed in the lower generation dendrimers. The similar redox potentials obtained under various solvent conditions indicate that the increased shielding in the higher generation dendrimers creates a microenvironment in the core, increasingly inert towards changes in the surroundings.[10]

Dendrimeric zinc-porphyrin complexes mimicking natural zinc-porphyrins show generation-dependent electronic behaviour. The mimics comprise carboxylic acid-terminated Fréchet-type dendrons bound to the zinc-porphyrin core. While the dendrimers have similar electronic properties (according to their absorption spectra) in apolar solvents like dichloromethane, a strongly generation-dependent behaviour is observed in polar solvents. Upon increasing the ionic strength or lowering the pH resulting in a more "hydrophobic behaviour" of the dendritic porphyrin a decrease in the strength of the porphyrin-specific transition (Soret transition) was observed. This decrease in absorbance (hypochromicity) may be the result of a contraction of the dendritic structure being increasingly hydrophobic relative to the surroundings, resulting in increased energy transfer between the porphyrin and the aromatic residues in the more contracted dendritic structure. Also, in interactions with methyl viologen (MV), a commonly used positively charged electron acceptor and with negatively charged electron acceptors (*e.g.* naphthalenesulfonate) in polar solvents, the high-generation dendritic porphyrins show an almost inert behaviour upon change in acceptor concentration. In comparison, the lower generation porphyrins with their more open structure, show a better ability to interact with both positive and negatively charged acceptors.[11]

PAMAM dendrimers with peptide-modified surfaces, capable of coordinating multiple zinc mesoporphyrins, have been applied as electron transporters in the enzymatic formation of hydrogen in artificial photosynthetic systems (Figure 6.5).[12,13] In contrast to the above-mentioned electron transport mimics, the present mimics have multivalent zinc-porphyrin systems at their surfaces with a valence proportional to the dendrimer generation. As consequence, it was found that the reduction of MV was highly generation dependent, with increased reductive power upon increasing generation. Electrostatic interactions between the dendrimer and MV proved to be an important factor for the efficacy of single-electron transfer (S.E.T.) between the dendrimer and MV. This electronic transfer is indicated by the formation of the blue MV radical cation. As a consequence, the use of anionic PAMAM dendrimers gave poorer electron transfer rates compared to cationic dendrimers, as the positively charged MV was bound to the negatively charged dendrimer, thereby increasing the "back electron-transfer" from MV to the dendrimer.

Also, non-porphyrin-based natural systems capable of electron transport such as ferrodoxin have been mimicked by dendrimeric models based on an inorganic iron–sulfur cluster core $[Fe_4S_4(SR)_4]^{2-}$ encapsulated by Fréchet-type dendrons.[14] The reversible single-electron transfer in these iron–sulfur clusters are well investigated. Like the haem-containing systems, these systems show a generation-dependent behaviour in electron transfer processes, whereas high-generation systems show the strongest deviation in electronic properties from the unmodified core. The reduction

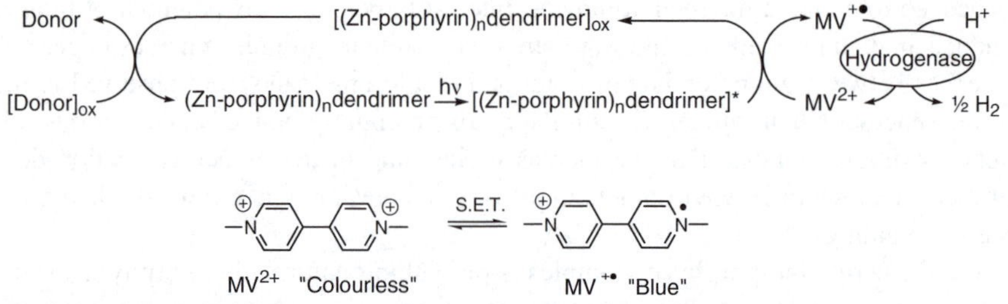

Figure 6.5 *Hydrogenase catalysed hydrogen formation, involving a dendrimer in the electron transfer cycle. Donor: Triethanol amine. MV: Methyl viologen. SET: Single-electron transfer*[12]

of the core becomes increasingly difficult upon increasing generation as a result of the increased shielding from the surroundings provided by the dendritic wedges.

Dendrimeric oxygen-transporters: Haem-containing proteins are, in addition to their function as electron transporters, used for the storage and transport of molecular oxygen in many aerobic organisms. In proteins like haemoglobin or myoglobin, π-electron donation from the molecular oxygen ligand results in an Fe(II) oxygen complex. This low oxidation state is stabilised by protein encapsulation from the redox-induced decomposition to peroxide and Fe(III). Nearby histidine imidazole residues prevent this complex from oxidative decomposition by hydrogen bonding. The introduction of a dendritic shell as a protein mimic opens up for facile well-defined modulation of, *e.g.* hydrogen bonding and shielding properties, which may give information on the factors influencing oxygen binding and oxidative decomposition of these complexes.[15] The favourable properties of dendrimer shells as protein mimics for haem cores have been predicted on theoretical basis using computer simulations (molecular mechanics and molecular dynamics). The calculations conclude that haem cores encapsulated in G1–G5 dendrons all are capable of forming stable complexes with molecular oxygen, with increased stability for the G5 adduct.[16] "Globin" mimics based on PEG-terminated polar amide dendrons as molecular shields, have been shown to form stable complexes with molecular oxygen (Figure 6.6).[17] This system did not show any significant generation-dependent increase in the ability to bind either oxygen or carbon monoxide. However, the ability to form hydrogen bonds to the oxygen ligand proved to be of great importance. A strong decrease in complex stability was observed by exchanging the amide-containing dendrons with dendrons containing ester moieties. In contrast to unsubstituted amides, the ester functionalities are not capable of acting as hydrogen donors for hydrogen bond interaction with the oxygen ligand. Also, the polarity of the solvent has a great influence on the stability of the iron-porphyrin–oxygen complexes. Applying water as solvent destabilises the complexes by disruption of the oxygen–amide hydrogen bonds. In the dendrimeric systems, the binding of carbon monoxide, although, being a π-acceptor ligand capable of forming strong complexes with iron preferentially to oxygen, was hampered by sensitivity to steric congestion in the binding site. As a consequence preferential binding of oxygen relative to carbon monoxide was found in these artificial Hb systems. Previous reported systems

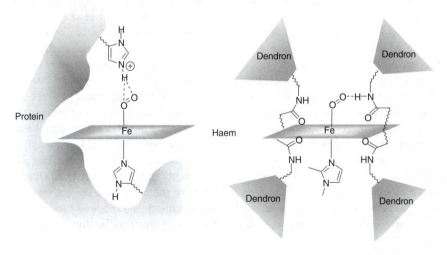

Figure 6.6 *Stabilisation of oxygen binding, by hydrogen bonding in, left: haemoglobin and right: dendritic haemoglobin mimic with loosely bound dimethyl imidazole as axial ligand*

of similar structure containing a tethered axial imidazole similarly showed good binding properties of oxygen, and were in addition to the binding of oxygen and carbon monoxide found to bind NO.[18] In the natural haem proteins, the lack of sterical congestion results in a preferential binding of carbon monoxide, since the binding affinity in these systems are more governed by electronic factors.

Cross-linking of haemoglobin into clusters, may mimic the very high local concentrations of haemoglobin found in red blood cells, create powerful semi-synthetic oxygen carriers, potentially useful as red blood cell substitutes in short-term treatment accompanying blood transfusions. Dendrimers (G4-PAMAM) have been used as "core" in high molecular weight (110–142 kDa) dendritic clusters having tetrameric haemoglobin as terminal groups. Conjugation of tetrameric haemoglobin to the dendrimer resulted in an increased oxygen-affinity accompanied by a reduced cooperativity in oxygen uptake of the clusters in comparison to conventionally cross-linked haemoglobin.[19] In haemoglobin this cooperative effect leads to an increased oxygen uptake upon binding of oxygen to one of the haemoglobin subunits. This cooperativity is a result of the changed geometry of the iron complexes upon binding of oxygen leading to subsequent rearrangement of the protein structure. The reduced cooperativity in the clusters is attributed to the increased amount of haemoglobin–haemoglobin interaction found in the clusters, in which the higher sterical congestion may hamper the conformational induced changes by oxygen uptake and release.

6.3 Dendrimers as Artificial Enzymes

Although, the ability of dendrimers to mimic proteins is interesting in connection with development of protein mimic-based biomaterials or transport vehicles, further modification of the dendritic molecules may take the protein mimicry a step further – *e.g.* in the creation of artificial enzymes. In nature, the action of enzymes on a molecular level is a highly complex matter involving multiple factors such as the spatial arrangements of the peptide chain and side-chains with intrinsic electronic properties (*i.e.* electron donor–acceptor). The well-defined spatial nature of the enzyme active site leads to phenomena such as anchimeric assistance and fixation of the reactants, which direct the enzyme towards specific substrates and specific modes of action. In the active site of the enzyme these factors reduces the transition state energy (lower activation energy) and consequently facilitates a particular reaction sequence on the substrate. In addition, the hydrophilicity or hydrophobicity of the active site may have a favourable effect in selective binding of substrates with similar polarity, and repel substrates with different polarity. Conformational changes in the protein as a consequence of participation of the active site in catalysis may give rise to cooperative effects, increasing the specificity and function further.

The interaction of an enzyme with a substrate may, roughly be divided into two parts:

- The "hosting part" – the ability of the enzyme, via specifically shaped cavities and surface polarity potentials, to select a specifically shaped substrate, *e.g.* being able to distinguish between enantiomers, epimers or regioisomers and polarity of a certain substrate.
- The "action part" – the ability to perform certain chemical transformations, when the properly shaped substrate is taken in.

At the present time, the reported number of synthetic molecules capable of both these modes of action are sparse, however, numerous investigations on these two fields separately is an ongoing exciting task. Molecular imprinting is an interesting tool in the creation of "passive" receptors capable of differentiating between different molecular shapes emulating the "hosting process" of enzymes and receptors found in nature. Dendrimers have predominantly been used in the "action part" of enzyme mimics, although, examples of dendrimers used to create the hosting part of an enzyme mimic by molecular imprinting are known (*vide supra*).[7] However, in synthetic chemistry, catalysts in general do not have the ability of discriminating between subtle differences in the molecular shapes of possible substrates. The use of enzymes, or genetically modified "substrate tolerant" enzymes is, and will be very useful tools in organic synthesis in years to come.

Although enzymes are highly complex molecular machines designed to act on certain substrates in a highly specific way, the use of dendrimers as models, although simplified, may give an insight into one of the intrinsic properties of enzymes, namely, the site isolation from the surroundings. The microenvironment created inside higher generation dendrimers may, in the case of dendrimers having polar scaffolds, create a polar interior even when the solvent surroundings are apolar. One such example was reported by Fréchet's group,[20] who used a lipid-decorated Fréchet-type dendrimer modified with a polar hydroxymethylated scaffold as catalyst for the unimolecular elimination of tertiary alkyl halides to alkenes in apolar media. This system with its apolar surface and polar interior resembles the built up of a micelle.

Although providing a greatly simplified picture of "enzymatic" activity, several properties of this dendrimer resemble the action of an enzyme. The alkyl halide, being polarisable is transported into the dendrimer interior, where the increased polarity of the core polarises the reactant, and serve as driving force for the transport of the reactant into the core. The adjacent oxygene lone-pairs of the dendrimer provide stabilisation of the cationic E1 transition state (T.S.) facilitating the formation of the alkene (Figure 6.7). The alkene is then, due to its low polarity, transported out of the core of the dendrimer into the apolar surrounding solvent. High conversion of the alkyl halide substrates (90–99%) was achieved in the presence of less than 0.01 mol% dendrimer catalyst. Control experiments without presence of dendrimer, resulted in no, or very low yields of the alkene.

In addition to the catalysis of E1- reactions, these dendrimers were found to catalyse S_N2-type bimolecular substitution reactions between methyl iodide and pyridine, leading to the N-methylated pyridinium compound. Like the E1 elimination reactions, the S_N2 reactions proceed through a polar transition state, which is similarly stabilised by the polar dendritic interior. The E1 and S_N2 catalytic activity of these modified Fréchet-type dendrimers agrees well with investigations on polarity using a *p*-nitroaniline solvatochromic probe, where the polarity of the surroundings affects the spectral properties of the probe.[21] Correlation between these investigations and Taft's solvent polarisability parameter[22] shows that the interior of a Fréchet-type dendrimer has a polarity similar to that of dimethyl formamide (DMF), which is also a common solvent for facilitating reactions involving polar transition states.

Another enzyme mimic that also takes advantage of a specific type of interaction between the substrate and the dendritic interior is bis-selenium-containing glutathione

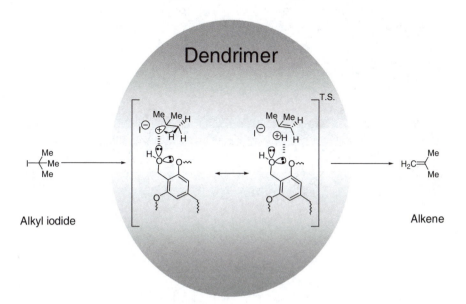

Figure 6.7 *Dendrimer catalysing the E1-elimination of tertiary alkyl halides under the formation of alkenes*

peroxidase (GPx) mimic (Figure 6.8). GPx is a seleno-containing enzyme present in mammalian organisms, which protects a large variety of cellular systems from the oxidative damage by catalysing the reduction of peroxides to the corresponding hydroxyl compounds. In an artificial biosystem based on a bis-arylseleno core encapsulated by Fréchet-type dendrons, glutathione has been substituted with thiophenol as the reductant. The catalytic reduction of hydrogen peroxide to water in presence of dendrimer was catalysed in a generation-dependent way with increasing catalytic activity upon increasing generation. The G3 dendrimer (1 mol%) gave an approximately 3000 times increase in reduction rate in comparison to the system without catalyst.[23] It is speculated that the initial intake of thiophenol into the dendritic interior may be favoured by hydrophobic (π-interactions) between the aryl-containing Fréchet dendrons and the aromatic thiophenol. This postulate was further confirmed by the ability of the G3-dendrimer to form host–guest complexes with thiophenol ($K_{ass} > 252.7 \text{ M}^{-1}$).

The best conversion rates for this system was obtained in polar solvent mixtures, where the poor ability of the solvent to solvate the dendrimer catalyst result in a more globular contracted conformation of the dendrimer, with higher shielding of the catalytic site, and more favourable intake of thiophenol.

Breslow's group[24] used PAMAM dendrimers with a pyridoxal core in an attempt to catalyse transamination, decarboxylation and epimerising reactions. Whereas the catalytic ability of these systems in transamination or decarboxylation was not clearly proven, these mimics showed to be good catalysts for the base-induced racemisation

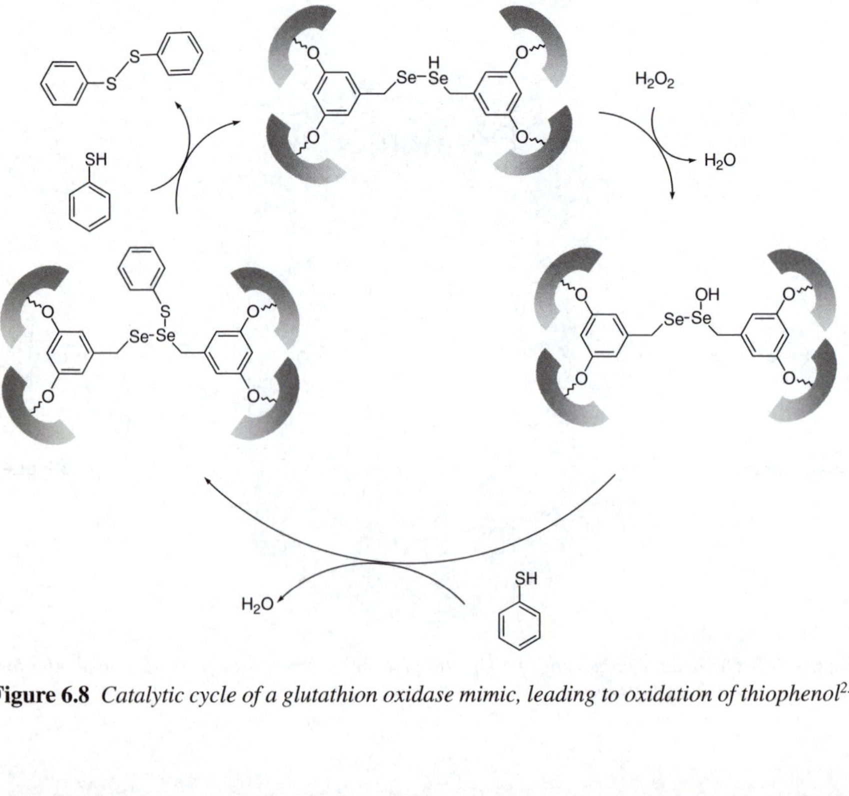

Figure 6.8 *Catalytic cycle of a glutathion oxidase mimic, leading to oxidation of thiophenol*[23]

of amino acids (50–100 times rate increase) bound to the pyridoxal core via an imine bond. The PAMAM scaffold, with its basic amines, serves as an ideal scaffold for this base-catalysed reaction (Figure 6.9). Though not shown in these studies, these pyridoxal catalysts may be useful in the post-modification of polypeptide chains obtained, *e.g.* by tryptic protein degradation or gene-expression because of their ability to specifically catalyse L-D epimerisation of the N-terminal residue in the peptide chain. Substitution of the peptide chain with one or a few D-amino acids often leads to greatly enhanced stability towards enzymatic degradation *in vivo*, which is an important tool in the development of peptide drugs. On the other hand, the biological activity may also depend on a specific stereochemical interaction, so the presence of D-amino acids in peptide drugs may quench the drug activity of the peptide.

One class of enzymes that has been subject to intense investigation is the proteases which catalyses the hydrolytic cleavage of peptide (amide) and ester bonds to the corresponding acid and amine or alcohol. These enzymes are an important factor in the digestion of proteins by tryptic degradation in the organism. The large substrate tolerance of some proteases make them particularly useful as alternative reagents in

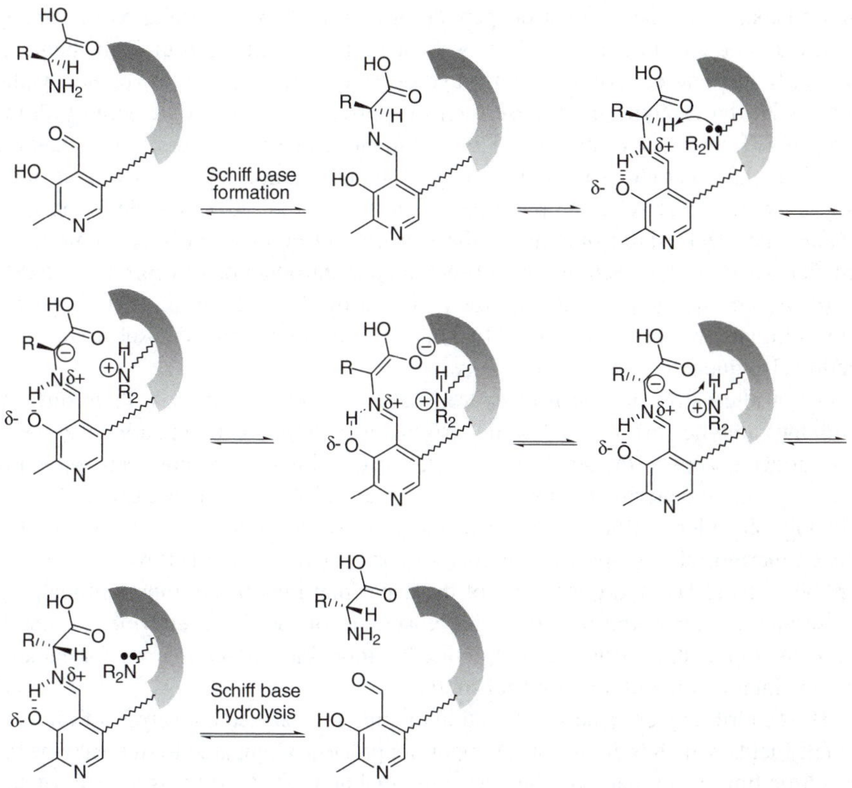

Figure 6.9 *Racemisation of an amino acid by the action of a PAMAM dendrimer (shown as a grey arch) with a pyridoxal core. The focal tertiary amines act as base catalyst for the racemisation of the pyridoxal-activated imino acid derivative*

synthetic organic chemistry. In addition to their hydrolytic activity under natural conditions, the enzymatic activity of these enzymes may be reversed to form amide bonds from the corresponding carboxylic acid and amine by careful control of the solvent conditions. It was soon evident that the catalytic activity of this class of enzymes relies on the presence of a catalytic triad in the active site consisting of serine, histidine and aspartic acid, and several synthetic and semi-synthetic compounds have been utilised as carriers for multivalent presentation of this tripeptide sequence, in order to create artificial proteases.[25] Recently, dendrimers comprising the catalytic triad have been synthesised using 3,5-diamino benzoic acid as the branching unit.[26] These artificial proteases showed catalytic activity in hydrolysing a variety of "active" ester substrates (active esters are particular electrophilic ester substrates) following Michaelis-Menten kinetics with constants in the order of $K_{\text{Michaelis-Menten}}$ ~0.1–0.55 mM. Histidine plays a dominant role in the catalytic ester cleavage, possibly by participation of *his*-imidazole as nucleophilic catalyst, as well as electrostatic stabilisation of the oxoanion intermediate by protonated imidazole residue. Aspartic acid similarly plays a role in the catalysis, albeit to a much lesser extent. Aspartic acid may have a role in the electrostatic attraction of the cationic quinolinium ester substrates, however, no clear indication was found in this regard. The catalytic capacity of these dendritic peptides was found to be similar to previously reported protease mimics based on, *e.g.* histidine containing four helix bundles. Similarly, peptide dendrimers with a hydrophobic core (*e.g.* alanine or phenylalanine) with the catalytic triad present on the surface, show catalytic activity in the hydrolytic breakdown of pyrene-1,3,6-trisulfonate-containing esters.[27] Also here the histidine plays a crucial role in the catalysis acting as a nucleophile as well as binding of the negatively charged pyrene esters to the protonised imidazole in the sidechain. The electrostatic binding of the substrate prior to esterolysis seems to be crucial, as no catalytic activity was observed upon introduction of neutral substrates. Furthermore, the enzyme activity was inhibited by the addition of 1,3,6,8-pyrenetetrasulfonate, indicating that the blocking of electrostatic enzyme–substrate interaction also quenches enzyme reactivity.[27]

As an alternative approach to hydrolytic dendrimers modified at their surfaces with the catalytic part of a hydrolytic enzyme, is conjugation of an entire enzyme to the surface groups of the dendrimer (Figure 6.10). This strategy has been applied in the synthesis of dendrimeric lipase enzyme mimics.[28] As dendrimeric carrier G2 and G3 poly-phenylene sulfide (PPS) dendrimers were utilised due to its high hydrophobicity, thermostability and the presence of easily derivatisable carboxylic moieties on the surface. The hydrophobicity of the dendrimer may be an important factor in preserving the structural integrity, and bioactivity of the lipase enzyme, as lipases have an amphipathic structure comprising hydrophobic surface regions for enhancing the interaction with the lipid substrates.

The dendrimeric enzyme was tested in the lipase-catalysed hydrolysis of olive oil to fatty acids, which is an industrial important process. Compared to free enzyme the dendritic lipase had markedly increased thermal and pH stability, as a result of stabilising intramolecular protein–protein interactions. As a consequence the catalytic activity remained very high (98%) even after numerous catalytic cycles, and over a wide pH and temperature range. The strategy of making dendrimeric enzyme clusters

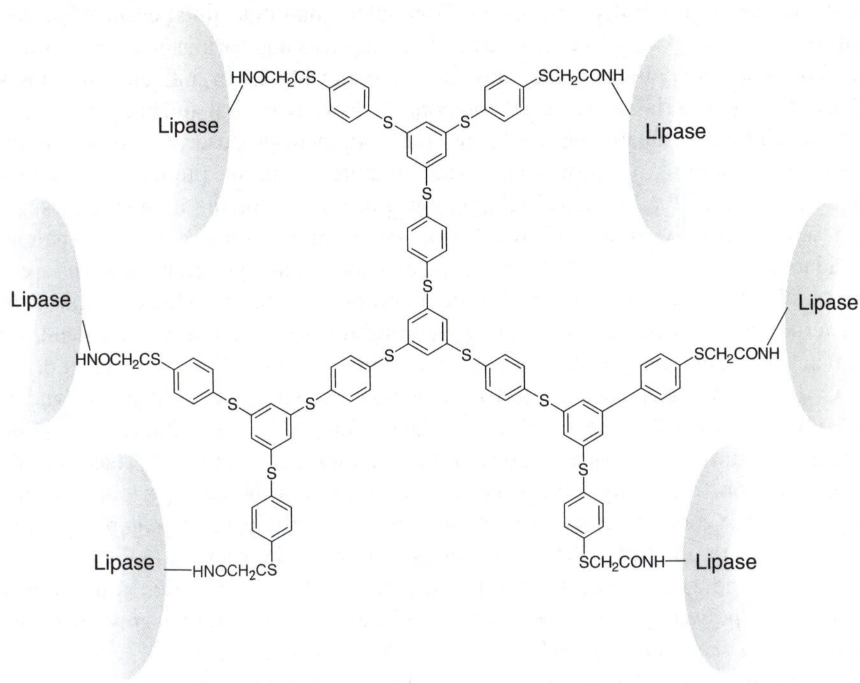

Figure 6.10 *PPS-dendrimer with its surface groups modified by multiple lipase enzymes, creating a highly reactive dendritic enzyme[28]*

may be useful in industrial applications, where the demand for enhanced enzymatic stability and durability is high.

PPI dendrimers surface modified with tungsten cyclopentadienyl complexes show photo-induced nuclease activity towards double stranded plasmid DNA *in vitro*. The basic PPI scaffold is a well-suited scaffold for initial electrostatic binding of the negatively charged DNA backbone under physiological conditions. The dendrimeric compounds showed higher efficacy in cleaving double as well as single-stranded DNA compared to the monomeric compound. However, no clear generation effect was observed, due to precipitation of dendrimer-DNA aggregates.[29]

6.4 Dendrimers as Artificial Antigens, Cell Surfaces and Antibodies

As the use of dendrimers as scaffolds for subsequent surface modification gives easy access to macromolecular structures having properties similar to the monomeric surface groups, this methodology is highly useful for the creation of molecularly simplified mimics of antigens, which with their complicated and sometimes highly irregular molecular structures may otherwise be hard to synthesise.

In nature, carbohydrates in the form of poly- and oligosaccharides often bound as glycoconjugates to proteins constitute an important class of compounds involved in numerous biological processes, ranging from inter- and intracellular events to recognition processes taking place between cellular surfaces and various microbes during infections. Carbohydrates being important building blocks in antigenic structures derived from bacteria and bacterial components, as well as viral and bacterial recognition points, are valuable targets for drug development purposes. Oligo- and polysaccharide mimicking bioisosters are valuable tools in pursuing a deeper understanding of the biological recognition processes involving carbohydrates. As we saw previously, numerous reports on dendrimers coated with carbohydrate residues on their surfaces show that these constructs have good antiviral and antibacterial activity (see Chapter 4). The function of these drugs, which interact with bacterial surface lectins or viral haemagglutinins, relies on the ability of the dendritic structure to mimic the cellular surface, thus acting as a competing artificial cell.

Carbohydrates have been utilised as branching core structure in the facile synthesis of oligosaccharide mimics, which can be derivatised further with, *e.g.* a peptide chain to create glycopeptide or glycoprotein mimicking structures.[30] These simple molecular constructs may serve as convenient mimics of the glycoprotein antennae residing at the cellular surface, which are important participants in, *e.g.* recognition processes with multiple lectins (adhesins) on bacteria prior to infection (see Chapters 2 and 4). In contrast to the tedious methodologies applied in conventional synthesis of polysaccharides, these building blocks are synthesised in good yields by applying chemoselective peptide coupling strategies (Figure 6.11).

A similar strategy has been applied in the synthesis of "pure" oligosaccharide mimics from branched mannosyl dendrimers containing triple antennae mannosylic residues centred around a trivalent trimesic core.[31] These procedures may be important tools in the creation of otherwise complex oligosaccharides and glycoconjugates with biological activity.

Alternatively, polysaccharide mimics can be created from commercially available scaffolds, *e.g.* PAMAM dendrimers. In these designs the inter-sugar molecular distances are modulated in a single final conjugation step between the high number of

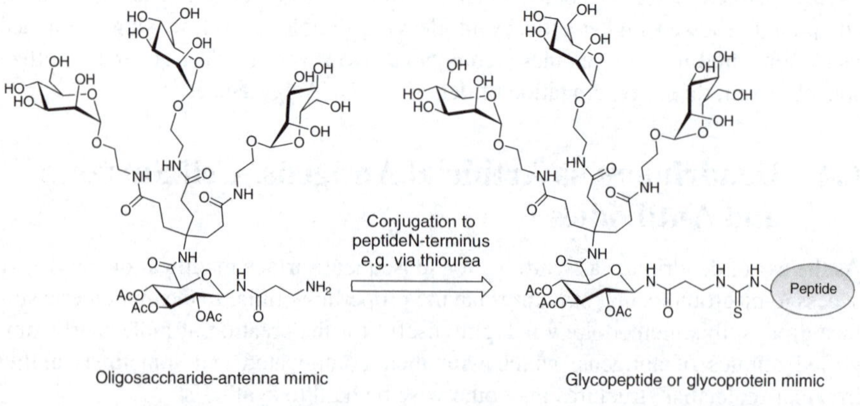

Oligosaccharide-antenna mimic   Glycopeptide or glycoprotein mimic

Figure 6.11 *Synthesis of glycopeptide or glycoprotein mimics by the use of oligosaccharide mimics*[28]

surface functionalities on the dendrimer and the modified carbohydrate, however, only highly symmetrical constructs can be made by this strategy. As the reaction between the carbohydrate (or preformed carbohydrate cluster) has to take place under mild and aqueous conditions, several methodologies from the well-established glycoconjugate field are taken into use, *e.g.* chemoselective amide, thiourea, urea formation by the use of active esters, N-carboxyanhydrides,[32] isothiocyanates or "active carbamates".

Although there are numerous reports on the use of glycodendrimers as artificial cell surfaces in order to inhibit microbial invasion, utilising carbohydrates normally found on cell surfaces, dendrimers covered with carbohydrates found on, *e.g.* cancer cells may be of great interest in diagnostics and therapeutics, *e.g.* in vaccine development (see Chapter 4). T-antigen, a disaccharide [β-D-Gal (1-3)-α-D-GalNAc] which is over-expressed and closely associated with malignant carcinoma cells as a result of aberrant glycosylations.[33] Glycodendrimers bearing this antigen serve as artificial densely carbohydrate-exposing cancer cells, which may be useful derivatives for the raising of T-antigen specific poly- and monoclonal antibodies, or by applying the appropriate immuno-modulators may have vaccine properties (Chapter 4). PAMAM dendrimers (G1–G3) have been applied as macromolecular carrier protein mimics for carrying penicillin.[34] The multipenicilloylated dendrimers may act as artificial multivalent antigens. By utilising dendrimers as carrier molecules a higher density and well-defined number of haptens can be carried per molecule in comparison to traditional carrier proteins, *e.g.* human serum albumin. Another disadvantage of using proteins as carriers for haptens is the resulting heterogeneous derivatisation of the protein as the reactive groups on the protein are heterogeneously located. Synthetic linear polymers for carrying a high number of haptens, *e.g.* poly-lysine do not have a well-defined (*i.e.* monodisperse) structure, and by conjugation, a mixture of conjugates may be obtained. In addition, the coiling and folding of linear polymers give rise to an infinite number of three-dimensional structures.

As penicillin with its β-lactam structure, is sensitive to nucleophilic attack, this sensitivity was applied in the conjugation with the surface amines of the denrimer, thereby creating a dendrimer with numerous pseudo-penicillin structures at its surface (Figure 6.12)

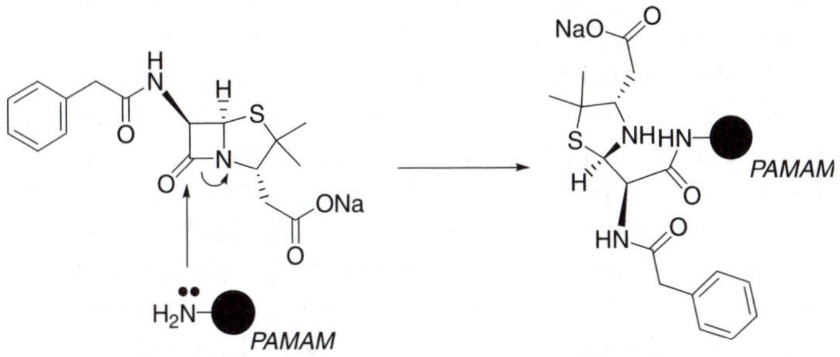

Figure 6.12 *Mild and facile methodology for the penicilloylation of an amino-terminated dendrimer by a β-lactam ring-opening amide bond formation, to give a dendrimer surface modified with numerous pseudo-penicillins*

Interestingly, in this case, a complete surface modification of a G3-PAMAM was facilitated at low temperatures, in contrast to the conventional thinking of heating the mixture in order to drive the reaction to completion. The dendrimers were tested as multivalent antigens compared to monomeric penicillin, which were reacted with butyl amine for the formation of a ring-opened pseudo-penicillin structure similar to the structures exposed on the dendrimer surface. The relative dendrimer–antibody binding affinities (IgE-antibodies) showed an increase in affinities with increasing dendrimer generation, the G3-PAMAM-penicillin adduct having a 100-fold increase in IgE affinity compared to the monomeric penicillin compound. This pronounced increase in binding affinities may indicate a dendritic or synergistic effect upon increasing the generation of the dendrimer, and may also increase the specificity of antibody recognition process.

G2-dendrons, functionalised with multiple aldols at their surfaces are found to be substrates for the antibody 38C2 catalysing the retro-aldol reaction under the formation of a dendron covered with multiple aldehyde functionalities.[35] In this case the dendron acts as an artificial antigenic substrate, which pave the way for highly biospecific recognition and subsequent degeneration of aldol-containing dendrimers, *e.g.* for the liberation of drugs from dendrimeric drug delivery devices (see Chapter 3).

6.5 Summary

Dendrimers are versatile tools for the creation of mimics of the great variety of functional molecules found in nature. The utility of dendrimers in their mimicry of collagen-like proteins rely on their dense (relative to *e.g.* proteins) well-defined structure, facilitating and stabilising the formation of tertiary and quartenary structures, which may lead to higher mechanical strength of these materials compared to their natural counterparts. The dense dendritic cage found in the higher generation dendrimers encapsulates the core region to create a microenvironment with properties different from the surroundings. This encapsulation effect is highly useful for the mimicry of enzymes, where one important function of the protein shell around the active site is to provide site isolation from the surroundings. In nature, however, the protein shell plays a dual role, in that the protein shell furthermore has a shape and polarity surface that fits the substrate. Also in the formation of substrate-specific molecular pockets, dendrimers can play an important role, as dendrimers formed around a core structure, due to their dense structure, creates a well-defined imprinted pocket of the core. The molecular imprinting methodology implemented by Zimmerman's group, may open up for the possibility, not only to make "passive" receptors for various substrates, but also in, *e.g.* combination with formation of basic, or hydrogen bonding groups, to facilitate, *e.g.* E1- or S_N2 reactions on substrates having certain three-dimensional molecular shapes, thus creating enzyme mimics that more closely resemble the substrate recognition and subsequent transformation observed in nature. The multivalent surface of dendrimers can be modified with enzymes forming multivalent enzyme clusters, with higher robustness and activity in comparison with the corresponding natural enzymes. Dendritic haem compounds such as "globin" and cytochrome mimics show properties resembling their natural counterparts, accompanying their mimicry of the oxygen and electron-transport properties. Furthermore, these dendrimeric haem-containing systems are beginning to constitute a new class of macromolecular reagents utilised in organic synthesis.[36]

Dendrimers covered with carbohydrates present on the cellular surface, may act as mimics of oligosaccharide antennae or act as artificial cells. As we have seen in Chapter 4, the ability to mimic the cellular surface, make these dendritic reagents well suited as drugs for the inhibition of the initial adhesion processes taking place prior to bacterial and viral infections.

In contrast to natural systems, the synthetic dendrimeric systems, which emulate the action of a given biomolecule, are generally more stable with less conformational responsive behaviour towards changes in the surroundings.

References

1. J.P. Tam, *Proc. Natl. Acad. Sci. USA*, 1988, **85**, 5409.
2. J.P. Tam and J.C. Spetzler, *Biomed. Pept. Proteins Nucleic acids*, 1995, **3**, 123.
3. L. Bracci, C. Falciani, B. Lelli, L. Lozzi, Y. Runci, A. Pini, M.G. De Montis, A. Tagliamonte and P. Neri, *J. Biol. Chem.*, 2003, **278**, 46590.
4. D.A. Tomalia, B. Huang, D.R. Swanson, H.M. Brothers II and J.W. Klimash, *Tetrahedron*, 2003, **59**, 3799.
5. M. Goodman, W. Cai and G.A. Kinberger, *Macromol. Symp.*, 2003, **201**, 223.
6. D.K. Smith, A. Zingg and F. Diederich, *Helv. Chim. Acta*, 1999, **82**, 1225.
7. S.C. Zimmerman, M.S. Wendland, N.A. Rakow, I. Zharov and K.S. Suslick, *Nature*, 2002, **418**, 399.
8. K. Haupt and K. Mosbach, *Trends Biotechnol.*, 1998, **16**, 468.
9. S. Kasai, H. Nagasawa, M. Shimamura, Y. Uto and H. Hori, *Bioorg. Med. Chem. Lett.*, 2002, **12**, 951.
10. P. Weyermann, J.-P. Gisselbrecht, C. Boudon, F. Diederich and M. Gross, *Angew. Chem. Int. Ed.*, 1999, **38**, 3215.
11. R. Sadamoto, N. Tomioka and T. Aida, *J. Am. Chem. Soc.*, 1996, **118**, 3978.
12. M. Sakamoto, T. Kamachi, I. Okura, A. Ueno and H. Mihara, *Biopolymers*, 2001, **59**, 103.
13. M. Sakamoto, A. Ueno and H. Mihara, *Chem. Eur. J.*, 2001, **7**, 2449.
14. C.B. Gorman, B.L. Parkhurst, W.Y. Su and K.-Y. Chen, *J. Am. Chem. Soc.*, 1997, **119**, 1141.
15. F. Diederich, *Chimia*, 2001, **55**, 821.
16. M. Fermeglia, M. Ferrone and S. Pricl, *Bioorg. Med. Chem.*, 2002, **10**, 2471.
17. A. Zingg, B. Felber, V. Gramlich, L. Fu, J.P. Collman and F. Diederich, *Helv. Chim. Acta*, 2002, **85**, 333.
18. P. Weyermann and F. Diederich, *J. Chem. Soc. Perkin Trans. 1*, 2000, 4231.
19. R. Kluger and J. Zhang, *J. Am. Chem. Soc.*, 2003, **125**, 6070.
20. M.E. Piotti, F. Rivera, Jr., R. Bond, C.J. Hawker and J.M.J. Fréchet, *J. Am. Chem. Soc.*, 1999, **121**, 9471.
21. C.J. Hawker, K.L. wooley and J.M.J. Fréchet, *J. Am. Chem. Soc.*, 1993, **115**, 4375.
22. M.J. Kamlet, J.M. Abboud, M.H. Abraham and R.W. Taft, *J. Org. Chem.*, 1983, **48**, 2877.

23. X. Zhang, H. Xu, Z. Dong, Y. Wang, J. Liu and J. Shen, *J. Am. Chem. Soc.*, 2004, **126**, 10556.
24. L. Liu and R. Breslow, *Bioorg. Med. Chem. Lett.*, 2004, **12**, 3277.
25. B. Ekberg, L.I. Andersson and K. Mosbach, *Carbohydr. Res.*, 1989, **192**, 111.
26. C. Douat-Cassasus, T. Darbre and J.-L. Reymond, *J. Am. Chem. Soc.*, 2004, **126**, 7817.
27. A. Clouet, T. Darbre and J.-L. Reymond, *Adv. Synth. Catal.*, 2004, **346**, 1195.
28. O. Yemul and T. Imae, *Biomacromolecules*, 2005, **6**, 2809.
29. A.L. Hurley and D.L. Mohler, *Org. Lett.*, 2000, **2**, 2745.
30. N. Röeckendorf and T.K. Lindhorst, *J. Org. Chem.*, 2004, **69**, 4441.
31. A. Patel and T.K. Lindhorst, *J. Org. Chem.*, 2001, **66**, 2674.
32. K. Aoi, K. Tsutsumiuchi, A. Yamamoto and M. Okada, *Macromol. Rapid Commun.*, 1998, **19**, 5.
33. R. Roy and M.G. Baek, *J. Biotechnol.*, 2002, **90**, 291.
34. F. Sánchez-Sancho, E. Pérez-Inestrosa, R. Suau, C. Mayorga, M.J. Torres and M. Blanca, *Bioconjugate Chem.*, 2002, **13**, 647.
35. A. Córdova and K.D. Janda, *J. Am. Chem. Soc.*, 2001, **123**, 8248.
36. M. Uyemura and T. Aida, *Chem. Eur. J.*, 2003, **9**, 3492.

Subject Index

Aβ amyloid, 123
Adhesion, 91, 92, 93, 95, 96, 100, 101, 106, 171
Adjuvant, 57, 59, 107, 110–119, 124
ADP (recognition), 74, 75
Aggregate, protein, 111, 116, 119–123, 153
Allergic, 117, 141
Alzheimer's disease, 119, 123, 124, 153
2-Aminododecanoic acid, 84
2-Amino-tetradecanoic acid, 84
2-Aminooctadecanoic acid, 84
AMP (recognition), 74, 75
Amplified genealogically directed synthesis, 9
Aminoxyacetylation, 113
Amphipathy, 97
Amyloid, 121, 123, 124, 153
Angiogenesis, 118
Angiography, 132–133
Artificial enzymes, 153, 161, 162,
Artificial cell surfaces, 169
Arboroles, 3
Arborescent polymers, 3
Asymmetric dendrimers, 12, 14, 69
Antiadhesins, 31, 101
Antiangiogenesis, 157
Antibacterial activity, 67, 71
Antibacterial dendrimer, 97, 98, 100–103
Antibody, 106–109, 112, 113, 114, 116, 117, 120, 136–141, 144, 154, 170
Antibody array, 144
Antifungal activity, 67
Antitumour dendrimer, 103, 116
Antiviral dendrimer, 91, 92, 94, 95, 97, 103, 168
ATP (recognition), 74, 75

Autoimmunity, 97, 108, 10
Avidin, 140–141
AZT (binding of), 72, 73

BAC, 82
Back-folding, 18–24
Bacteriocide, 100
Barbituric acid (binding of), 72, 73, 78
Barium sulfate (contrast agent), 130
Barrel stave mechanism, 97
B-cell, 108, 109, 111, 114, 118
Bead, 138
Bilayer membrane, 100, 119
Binding assay, 92, 101
Bioassays, 135–149
Biosensor electrode, 141
Biotin, 138, 140, 141
Biodistribution, 63, 70
Biopermeability, 46, 47, 58
 -Transepithelial, 49, 51, 52, 54, 55, 59
 -Transendothelial, 52–56
 -Extravascular, 53
Bismuth, 132
Blood circulation time, 54, 56, 58, 81
Blood half-life, 56
Boltzman distribution, 130
Bovine spongiform encephalopathy, 119
"Bow tie" dendrimer, 69–70
Bromoacetylation, 113
1-Bromoacetyl-5-fluorouracil, 76

Caco-2 cells, 78
Camptothecin, 80
Cancer, 62
 -B16F10 melanoma, 70
 -MDA-MB-231 cells, 70
 -Molt-3 leukemia cells, 80
 -and camptothecin, 80

-and *cis*platin, 69
-and doxorubicin, 80
-and fluorouracil, 66
-and folate-receptor, 78–79
-and methotrexate, 79
-and paclitaxel, 80
-and PAMAM, 66, 69
Candida albicans, 106
Carbohydrate moieties, 115, 138
Carcinoma, 104, 106, 169
Carpet model, 97
Carrier protein, 108, 109, 114–116, 169
Cell lysis, 97
Chaperones, 15
Chaotrope, 123
Chemoselective coupling, 113, 168, 169
Chlamydia, 93
Cholera toxin, 102
Cholesterol, 81–82
*Cis*platin, 69
Cluster effect, 101, 117
Cobaltocene, 72
Coiled coil peptide, 122
Collagen mimic, 153
Complement, 108, 110, 119, 122,
Contrast agent, 91, 104
Controlled release, 62–63, 66, 71
Cooperative binding, 29
Coupling ratio, 109
Coupling orientation, 109
CpG, 119
Creutzfeldt Jakob's disease, 119, 153
Crohn's disease, 77
CT, 130, 131
Cy5, 146
Cyclodextrins, 72
Cytochrome P450, 73
Cytokine, 108, 109, 114, 116, 118
Cytomegalovirus, 96
Cytotoxicity, 32, 36–43, 45–58, 78,79, 83, 91–98

Dark toxicity, 103
DC-SIGN, 96
Defects, 10, 14, 83

Defensin, 97, 98, 100
De Gennes model, 18
Dendrigrafts, 3, 4, 96, 116, 138
Dendrimer conformation,
-effect of molecular growth, 17, 18
-effect of pH, 20
-effect of solvent, 21
-effect of salts, 23
-effect of concentration, 23
-extended, 20–23
-molecular simulations, 18, 19, 23
-"native", 17
-NMR, 19, 21, 22, 64, 68, 72, 76, 86, 130
-SANS, 21
-SAXS, 23, 65
-tight, 17
Dendrimer, DNA-based, 135
Dendrimer, light-harvesting, 146
Dendrimer, penicilloylated, 141, 169, 170
Dendrimer, phosphorus-containing, 121
Dendrimer drug, 90
Dendrimer structure
-Core, 21, 42, 63, 64, 71
-Focal point, 4, 6, 8
-Generation, 4–6
-Shell, 4, 5, 7, 16
-Surface groups, 4, 6, 7, 34, 36, 39, 40, 45–58
Dendrimer-DNA complex, 40, 41, 47, 48, 119, 167
Dendrimeric electron transporters, 157, 158
Dendritic box, 18, 70, 71, 75, 76
Dendritic cell, 92, 96
Dendritic effect, 29, 90, 106
Dendroclefts, 71, 72, 154
Dendron, 3, 4, 6, 8, 12
-amphiphilic, 100
Dendrophanes, 71, 72
Dense shell packing, 18, 19
Detection limit, 131, 144
1,12-Diaminododecane, 66, 67
Diastereoselectivity, 71
Diphenylhexatriene, 64

Disassembling dendrimer, 90
DMSO, 69
DNA, 62, 76, 141, 142
  -dendrimers, 135–138
  -extraction of, 147–148
  -and immobilization, 139–140
  -and microarrays, 141–147
  -and transfection, 81–85
  -and signal amplification, 135–138
DNA crosslinking, 144
DNA extraction, 147
Doxorubicin, 80
DOTAP, 83
Drug carrier, 90
Drug delivery, 90, 106, 107, 110, 153, 170
DTPA, 132, 133

Ebola, 92, 96
EDTA, 132
Efflux, 78
ELISA, 135–138
End-group, 5, 6, 8, 18, 21, 69, 74, 132
Endocytosis, 32, 36, 37, 41, 47, 48, 50, 51, 54, 63, 78, 81, 82, 85, 98
Endosome, 36, 37, 47, 81, 82
Enhanced permeability and retention effect, 63, 69, 85
Enterotoxin, 102
EPR-effect, 63, 69, 85
Epstein-Barr virus, 145
Escherichia coli, 101
Eukaryotic cells, 81

Ferrocene (and derivatives), 131–132, 139
Fimbriae, 101
FITC, 82, 134, 135
Flexibility, 106
Fluorescence, 64, 78–79, 82, 103, 133, 137, 143, 145, 146, 147, 149
Fluorescence resonance energy transfer, 133, 135, 145–146
Fluorescein, 64, 69, 82, 134, 145, 146
Fluoresceine isothiocyanate, 82, 134, 135

Fluorophore, 133, 134, 135, 138, 143–147
Fluoruracil, 66, 67, 76, 77
Folate, 78, 104
Folate receptor, 133, 145
Foot-and-mouth disease, 113
Fourier transform infrared spectroscopy, 138, 139
Fp-maleimide, 138
Frëchet-type dendrimer, definition, 6
Fragmented dendrimers, 42, 43, 48
FRET, 133, 135, 145, 146
Freund's adjuvant, 113, 114, 124
FTIR, 138, 139

Gadolinium, 130, 131, 132
Gadomer-17, 131, 133
Ganglioside, 102, 103
Gene therapy, 62, 81, 85
Genital herpes, 93
Glucose oxidase, 141, 142
Glucoside, 71
Glutarimide, 72, 73
Glycodendimer, 8, 31, 95, 96, 101–106, 115, 116, 169
Glycoprotein, 92, 96, 100, 101, 168
Glycoprotein-P, 78
Glycoside cluster effect, 29
Glycosphingolipid, 100
Gold surfaces, 135, 139–141
Gram negative, 97–100, 114
Gram positive, 71, 97–100

Haemolysis assay, 34
Haemolytic, 91
Haemolytic activity, 66
Haemagglutination assay, 95, 101
Haloacetylated amine, 113
Hamilton-receptor, 76
Hapten, 108, 112, 117, 118, 140, 141, 149, 169
Heat shock proteins, 15
Hepatitis, 114,
Herpes, 134, 145
Herpes simplex virus, 92, 96
Heterogenous assay, 139

HIV, 92–96, 107, 114
Hole formation, in membranes, 36, 41, 47, 50, 54, 55, 98
Horseradish peroxidase, 138
Host-guest complexes, 63–76
Hybridisation, 141–144
Hydrodynamic ratio, 66
Hydrogen-bonding, 12, 13, 15, 17, 18, 20, 22, 24, 64, 71, 72, 74, 76, 94, 160, 161
10-Hydroxycamptothecine, 68

Ibuprofen, 73
IC50, 92, 95
IgA, 114
IgE, 117, 131, 144
Immune response, 96, 107, 110, 112, 115, 117, 124
Immune system, innate, 97
Immune system, non-adaptive, 97
Immunogen, 106, 108, 109, 110, 114, 115, 116, 124
   -synthetic, 110, 114
Immunogenicity, 32, 57, 59, 90, 107–117, 124
Immunological memory, 106–108
Immunosensor, 140
Immunosuppression, 116
Immunity, 107–109, 113, 124
Indomethacine, 66, 68, 69
Infection, 92–97, 100, 106–108, 114, 121, 124, 168
Infectivity, 92, 97
Inflammation, 108, 110, 119
Influenza virus, 92, 94, 97, 117
Intracellular infection, 114
Intracellular transport, 37, 47
In vitro
   -allergy testing, 134
   -drug delivery, 66, 69, 73, 76,
In vivo
   -degradation of dendrimers, 63
   -detection of metalloproteases, 134
   -drug-delivery, 69, 77
Iodine, 130, 131
Iron, 130, 138

Iron carbonyl, 138
Iscom, 117
Isoelectric point, 21
Isothermal calorimetry (ITC), 85

Kidney, 66, 69, 79, 133

LDH assay, 34, 35
Lectin, 92, 96, 154, 168
Lectin, C-type, 92, 96
Leukemia, 80
Linear genealogically directed synthesis, 9
Lipase mimics, 166
Lipids, 81, 83, 84
Lipid bilayer, 98, 119
Lipopolysaccharide, 99, 114, 118
Listeria monocytogenes, 138
Liver, 69
Luciferase, 81, 83, 135, 137
Lung epithelial cells, 78
Lymphocyte proliferation, 112

Macrophage, 108
Magnetic resonance imaging (MRI), 104, 130, 131, 132
Magnetite (and particles), 131, 147, 148
Magnetite, 131, 147, 148
Major histocompatibility complex (MHC), 109
Malaria, 110
α-Mannose, 101
Maleimide, 138
Manganese, 130, 131, 132
MAP, 147
MAP, core, 112, 113, 118
MAP, definition, 6
MAP, lipid containing, 112
Mast cells, 117
Measles virus, 112
Melanoma, 70, 116
Melamine dendrimer, 36, 44
Membrane permeability, 148
Metal carbonyl, 138
Metalloprotease, 134
Methacrylamides, 67

Methotrexate, 68, 79
Methyl-prednisolone, 77, 78
Micelle, 64, 66
Micellanoic acid, 64, 65
Microarray, 133, 135, 136, 138,
    141–149
Microcalorimetry, 67
Microbicide, 93, 94
Minimal inhibitory concentration
    (MIC), 101
Molecular dimensions, 17, 50, 53, 54,
    55, 152
Molecular imprinting, 154, 155, 162,
    170
Molecular recognition, 72
MTT assay, 34, 40
Mucosal immunisation, 114
Multiple antigenic peptide (MAP),
    106, 110
Multivalent interactions, 28, 29, 33, 146

Naphthalene, 64, 65
Native peptide structure, 111
NEESA, 135, 137
Neoglycoprotein, 101
Neural cell adhesion molecule
    (NCAM), 106
Neuronal cell line, 120
p-Nitro-benzoic acid, 70
NMR (Nuclear magnetic resonance),
    64, 68, 71, 72, 76, 90
Nomenclature, 3, 7
Nuclease mimics, 167
Nucleotide extension and excision
    coupled signal amplification, 135,
    137

ODN-1, 83, 85
Oligonucleotide, 62, 76, 146
    -and dendrimer carrier, 83–85
    -and inhibition of VEGF, 83–85
    -and detection of herpes, 134–135
    -and dendrimer-enhanced signal
        generation, 135–139
    -and single nucleotide
        polymorphism (SNP), 143–145

Oligonucleotide tag, 138
Opthalmology, 69
Orthogonal protection, 113

^{32}P-labelling, 144
Paclitaxel, 80
PAMAM dendrimer, definition, 4,
    -as tool in assays, 140–149
    -covalent drug-conjugates, 76–79
    -and contrast-agents, 132–133
    -and fluorescence-enhancement,
        133–135, 137–139
    -and gene transfer (transfection),
        81–85
    -and host-guest chemistry, 66–71, 73
Paracellular transport, 50, 51, 52, 54,
    55, 56, 59
Pathogen, identification of, 138
PCR, 134, 136, 138, 141, 143–145, 149
PEGylation, 66
    -of surface groups, 40, 53, 55,
        57–59, 66
Penicillin G, 80
Peptides, 62, 71, 84–85, 134–135,
    146–147, 149
    -binding to dendrimers, 75–76
Peptide, antimicrobial, 97, 98, 101
Peptide, cyclic, 113
Peptide library, 147
Peptide specific antibodies, 110
Peripheral blood mononuclear cells, 145
Peroxidase, 141
Pertechnetate, 75
Phagocytosis, 76, 90
Pharmacokinetics, 132
Phenol blue, 64, 65
Phenylalanine, 70, 82
Phosphatidylglycerol, 98
Phospholipid, 97–99
Phosphoramidite, doubler, 144
Phosphoramidite, trebler, 144
Photodynamic therapy (PDT), 103, 133
Photosensitizers, 133
π-π stacking, 95
Pilocarpine, 69
Pinacyanol chloride, 64

Pincer, 4, 7, 74
Plasmodium falciparum, 113
Plasmon surface resonance
    spectroscopy, 135
Polyester dendrimer, 45, 96
Polyethyleneimine, 81–83
Polylysine, 81–82
    -dendrimers, 81
Polymerase chain reaction, 134, 136,
    138, 141, 143–145, 149
Polymers,
    -Linear, 4, 16, 37, 42, 53, 101, 120,
    169
    -Hyperbranched, 3, 4, 62, 100
    -Dendritic, 3, 4
Polyphenylene dendrimer, 116
Polyplexes, 81, 82, 148
Polypropylene imine (PPI) dendrimers,
    68, 140
    -contrast-agents, 131–133
    -host-guest chemistry, 64–66, 70–76
Porphyrin, 133, 134
Prion protein, 119
Primer, 134, 138, 143, 144
Prodrug, 103
Protease activity, 145, 146
Protease resistance, 118–120
Protein array, 138
Protoporphyrin, 104
Pseudomonas aeruginosa, 148
Pyrene, 66, 68

Rabbits, 69
Racemisation, dendrimer induced, 164,
    165
RAST, 141
Radioallergosorbent test (RAST), 141
Rats, 67, 77
γ-Rays, 133
Recognition molecule, 101
Relaxation, 130
Relaxivity, 133
Renal excretion, renal structures , 133
Respiratory syncytial virus (RSV), 94
Retrovirus, 96
Ribozyme, 106

RNA, 76
Rose Bengal, 64, 65, 70

SAC, 135, 136, 137
SA-PE, 138
Saponin adjuvant, 113
Scanning electron microscopy (SEM),
    36
Scanning tunnelling microscopy
    (STM), 140
Scintigraphy, 130, 131, 133, 134
Sendai virus, 95
Sensitivity, 133–138, 141f
Shiga toxin, 103
Sialic acid, 92, 94, 95, 96
Signal amplification cassette (SAC),
    135, 136, 137
Signal generation, 93, 135, 139–141
Signal-noise ratio, 144
Silicon based dendrimers, 7, 8, 20, 31,
    39, 103, 104
Silver ions, 71
Single nucleotide polymorphism
    (SNP), 143
Singlet oxygen, 103
Small Angle X-ray Scattering (SAXS),
    65
SNP, 143
Solid phase, 93, 135, 136, 139, 141,
    144
Solvent effect, 119, 122
9,9'-Spirobi[9H-fluorene], 71
Starburst dendrimer, definition, 4, 6
Streptococcal membrane protein, 114
Stereoids, 62, 71, 72, 77
Stoke's shifts, 145
Streptavidin, 138, 140, 143
Sulfonated naphthalene, 93
Superfect, 82, 83, 119, 121
Surface plasmon resonance (SPR),
    135, 139, 140
Synergy effect, 90
Synthesis, dendrimers
    -Amplified divergent, 10
    -Convergent, 4, 12, 110, 155
    -Divergent, 8–12, 31

-Liniar divergent, 10
-Self-assembly, 8, 11–15
SYTOX Green, 148, 149

T-antigen, 104, 106, 169
TAT-1 peptide, 107
T-cell, 108–112
T-cell epitope, 111–115
T_1 (Relaxation time, spin-lattice), 130, 132, 133
T_2 (Relaxation time, spin-spin), 130, 132, 133
Template-assisted synthetic peptide (TASP), 116
Tetanus toxoid, 114
1,4,7,10-Tetraazacyclododecane, 76, 77
Tetrachlorfluorescein, 65
Tetramethylrhodamine, 82, 134
Therapy, 103, 106, 133, 155
Thiol nucleophile, 113
Tiaconozole, 68
Tin, 132
TMR , 82, 134, 135
Tn-antigen, 115, 116
Toll-like receptor, 57, 118
Toxicity, 63, 69, 76, 78, 82, 83, 85
Toxin, 100–104, 123, 135
Transfection, 81–85
-agent, 106

Transition state (T.S.), 162, 163
Trans-epithelial transport, 46, 47
Trans-vascular transport, 46, 47
Transcellular transport, 46, 47, 50, 51, 54, 55, 59
Triazine dendrimer, 39, 45, 68, 69
Tripalmitate-S-glyceryl cysteine, 112, 114
Tropicamide, 69
Tumour, 133, 134, 145
-imaging, 104
-targeting, 104

Ulcerative colitis, 77
Unfolding, 121
Urea, 73–75
UV-VIS, 64, 65

V3 loop, 114
Vaccine, dendrimer-based, 107
Vaccine, synthetic, 110, 124
Vascular endothelial growth factor (VEGF), 83, 84, 155, 156
VEGF, 83, 84, 155, 156
Vibrio cholerae, 102
Viral replication, 94
Visualisation, 145

X-ray, 130, 131, 149